Oxford *Smart*

Key Stage 3 Science

Activate 3

Teacher handbook

Curriculum Editor: Dr Andrew Chandler-Grevatt

Alyssa Fox-Charles • Gillian Hush • Jo Locke • Josh Thomas

OXFORD UNIVERSITY PRESS

Contents

Chemistry 3 Unit Opener (*Cont.*)

Physics 3 Unit Opener 168

Introduction

About the series

The *Oxford Smart* Activate has been updated to align with the principles of the *Oxford Smart* Curriculum for Science. It is designed to match the Key Stage 3 Programme of Study, with careful consideration of the transitions from Key Stage 2 and onto Key Stage 4 to help prepare students for success at GCSE and equivalent qualifications.

Oxford Smart Activate is flexible to suit your preferred route through Key Stage 3. The Programme of Study is covered in *Oxford Smart* Activate 1 and 2, making it a perfect match for a two-year course. *Oxford Smart* Activate 3 offers consolidation and extension of core concepts through application of knowledge and practice in the skills needed for success at Key Stage 4.

Across the Student Books, Teacher Handbooks, and Kerboodle courses, *Oxford Smart* Activate allows you to track students' progress using reliable assessment and learning resources. Our expert author and editor teams all have teaching experience, and our curriculum editor is a school-assessment expert. You can be confident that *Oxford Smart* Activate provides the best support for your students and teachers.

About your Teacher Handbook

This Teacher Handbook aims to save you time and effort by offering lesson plans and guidance on differentiation and assessment with a page-by-page match to the Student Book.

Unit openers provide context for students and show the prior knowledge required for a topic at a glance. The chapter openers include tips for delivering the content and the checkpoint lessons support students who have yet to grasp a secure knowledge of the outcomes. Lesson plans are written for 55-minute lessons but are flexible and fully adaptable.

About the *Oxford Smart* Curriculum for Science

The *Oxford Smart* Curriculum for Science has been built to address what research shows to be the most pressing considerations for teachers and learners. It connects content, methods, assessment, insight, and evaluation to deliver a holistic learning experience. Using data, insights, and research, it equips teachers to deliver personalised learning for all students.

Designed with an explicit egalitarian purpose to enable teachers and students to experience the joy and wonder of learning, the *Oxford Smart* Curriculum for Science's intent is based on the following six pillars. More information on each pillar can be found in the CPD and Research hub on Kerboodle.

Six Pillars

Provide clear and coherent pathways

The well-researched progression and sequencing support transition from Key Stage 2, through Key Stage 3, and onto GCSE or equivalent qualifications. The pathways are described in the Curriculum narratives documents and embedded throughout *Oxford Smart* Activate. Unit and chapter openers clearly show the learning journey for students and highlight the common misconceptions identified by our research.

Coherence

Hold high expectations, aspirations, and ambitions for all learners

Oxford Smart Activate identifies strengths, opportunities for challenge, and addresses learning 'gaps'. Expectations are clearly defined using the Developing, Secure, and Extending assessment framework. Challenge and scaffolding are provided in the resources through opt-in, rather than task-based, differentiation.

High expectations

Enable responsive teaching and learning that continually evolves and improves

Regular opportunities for formative assessment are provided through in-book questions, Reactivate questions, mini-checkpoint tests, and checkpoint assessments, with embedded checkpoint follow-up lessons and recommended next steps allowing for easy intervention. Regular spaced practice is included to help students retain concepts.

Responsive teaching and learning

Support engaged, self-regulated, and metacognitive learning

The *Oxford Smart* Activate equips learners to succeed with the ability to understand and use learning strategies effectively, facilitating independence and resilience. Learners develop metacognitive skills with scaffolded resources and reflection plenaries for every lesson, while professional development and targeted resources support teachers in modelling metacognitive strategies and self-regulation in the classroom.

Metacognitive learning

Promote development of learner identity and identification with science

Oxford Smart Activate reflects diversity and promotes a sense of belonging for students. The Trigger interest activities and unit openers demonstrate where science impacts our world, enabling learners to make connections with other subjects and their everyday lives.

Learner identity

Stimulate fascination, awe, and wonder in discovery of self and the world around us

Oxford Smart Activate promotes the awe and wonder of science through a wide range of hands-on practical activities and hooks for each topic. The unit openers and Trigger interest sections of lessons provide suggestions to spark curiosity and amazement.

Awe and **wonder**

Route through Year 9

For the second edition of Activate, we wanted to overhaul our Year 9 offering to cater for the differing needs of schools following a two-and-a-half-year or three-year Key Stage 3 programme. *Oxford Smart* Activate 3 is a flexible provision that can be easily adapted by teachers to fit their needs or preferences. The Year 9 course comprises both a traditional book and Kerboodle approach or a practical projects approach. These can be used separately or in combination.

Oxford Smart Activate 3 Books and Kerboodle

The Key Stage 3 Programme of Study is fully covered in *Oxford Smart* Activate 1 and 2, perfect for schools offering a two-year Key Stage 3.

The focus of *Oxford Smart* Activate 3 is therefore on consolidation, application, and extension of the core concepts and skills from Activate 1 and 2. This will ensure that students are equipped with the Key Stage 3 knowledge and skills that they will need for success at GCSE.

Reactivate

Every chapter begins with a Reactivate lesson aimed at recapping prerequisite knowledge for the chapter from books 1 and 2, as well as ensuring that students are ready to begin applying that knowledge.

Apply

The next four or five lessons, the core of the chapter, are Apply lessons. In these lessons, students are asked to apply knowledge from *Oxford Smart* Activate Books 1 and 2 in new, unfamiliar contexts. These contexts may involve new applications of skills. The topics of these lessons have been carefully selected to cover the core concepts that students need to be confident with before progressing to Key Stage 4. Through these application lessons, students will gain experience that will help them to tackle AO2 and AO3-style questions and build the skills that will prepare them for extended response questions and required practicals at GCSE.

Reach

The last two lessons of the chapter are the Reach lessons (apart from chemistry chapter 3, where there is only one Reach lesson). These are intended to bridge the gap between Key Stages 3 and 4. Unlike the Apply spreads, they involve the introduction of new, Key Stage 4 content. The topics chosen are designed to clearly build on from the Apply lessons in the chapter.

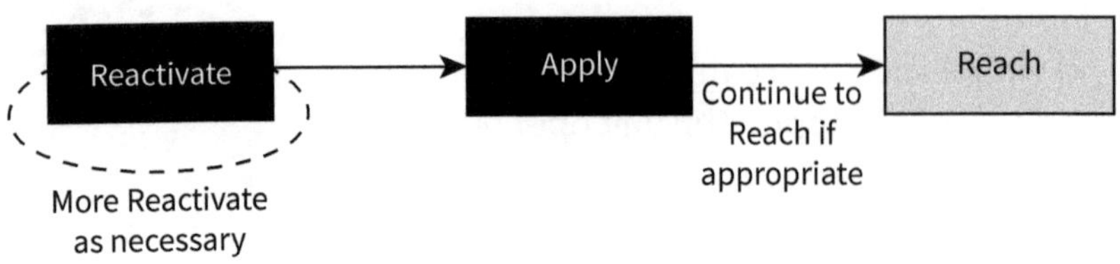

Adapting to your class

Data from mini-checkpoints and checkpoints from Years 7 and 8 could be used to inform how long you spend on the Reactivate and Reach portions of the chapter.

For some classes, one Reactivate lesson may be sufficient, others may benefit from a few lessons recapping, and some classes may only need an independent homework to review the content using the Student Book knowledge organiser and questions.

The Reach lessons may not be appropriate for all classes and you may choose to finish the topic after the Apply lessons. The checkpoint will assess only the Apply content to allow for this, with a separate Reach mini-checkpoint for those completing the Reach lessons.

Skills development

Skills development is an important part of *Oxford Smart* Activate and in Year 9, there is an opportunity to really ensure that your students are ready for the demand of GCSE. Metacognition, thinking and working scientifically, literacy, and mathematical skills are embedded throughout the course.

The Metacognition section in *Oxford Smart* Activate 3 focuses on effective revision strategies and exam technique, with guidance for teachers on how to introduce and develop these important skills.

The Working Scientifically chapter in the Student Book recaps the key information from Books 1 and 2 on planning investigations, recording and presenting data, and analysing and evaluating. The accompanying lessons in the Teacher handbook take students through a practical from start to finish to revise these skills and start to build independence. Further opportunities for this are built in throughout the course. The structure of these lessons and accompanying worksheets could easily be used to extend other practicals if you wished to.

Oxford Smart Activate 3 Practical Projects

An alternative option for part of Year 9 are the extended practical projects on Kerboodle. There are three cross-subject practical projects from which teachers can choose, each lasting approximately three weeks. They are designed to incorporate and focus on some of the core concepts that have been identified as important for students to grasp before Key Stage 4. The projects offer flexibility in terms of the amount of support that can be given to students. A teacher may complete one project in a more structured and guided form and a second project where the students have more independence. The projects will include lesson plans, worksheets, and other resources.

The projects could be interspersed with the book route, for example, they could replace similar topics or be used instead of the Reach lessons.

Research and CPD

The *Oxford Smart* Curriculum for Science and supporting resources and assessment in *Oxford Smart* Activate are based on a sound corpus of pedagogical research, and will generate its own evidence and impact from educators, experts, and real learning and assessment. Over time, this portfolio of evidence will in turn inform and inspire future curriculum evaluation and development.

What to expect from the research library

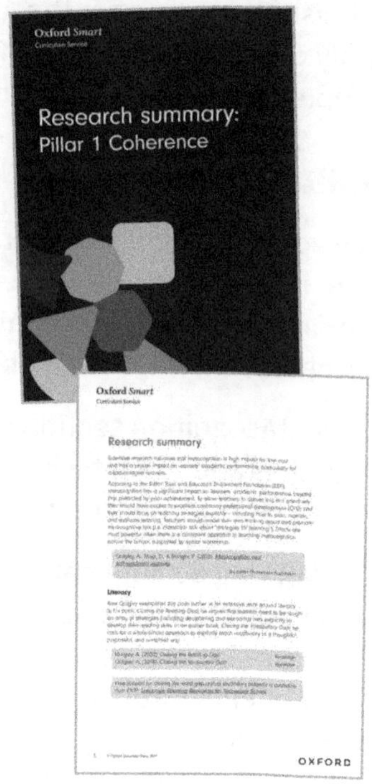

A Research library is available on Kerboodle. Here, we have collated the key pedagogical research and evidence that has been used to develop the *Oxford Smart* Curriculum for Science and *Oxford Smart* Activate, including a summary of the research that supports each pillar, references, and further reading.

In addition to the research used to shape the *Oxford Smart* Curriculum for Science are additional summaries of pedagogical literature that may impact the classroom now and in the future, including science-specific summaries and applications of the research.

The Research library will continue to be updated as the body of research grows to support the implementation of effective pedagogy in the classroom.

Impact

The *Oxford Smart* Activate has been trialled with 30 pioneer schools. Their feedback has informed the development of the curriculum, resources, and assessment.

Our pioneer schools have reported that some of the highlights of *Oxford Smart* Activate have been:

- the reduced planning time required for lessons
- the improved metacognitive practices from the embedded curriculum structure
- the enhanced student engagement from the range of lesson activities and practicals.

Continuing professional development

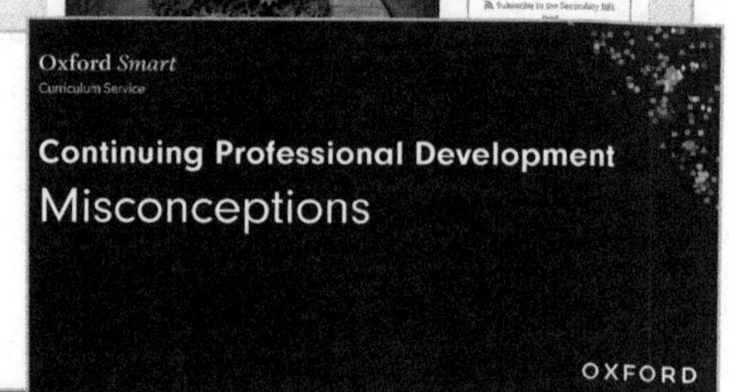

Alongside the Research library on Kerboodle, we are building a bank of CPD resources to support teachers as they implement *Oxford Smart* Activate.

These resources include blogs, articles, and videos that will cover:

- the six pillars of the *Oxford Smart* Curriculum for Science, how they are implemented in *Oxford Smart* Activate, and practical support for the classroom
- topic-focused support to help teach tricky topics or to approach a new topic for ECT and non-specialists
- targeted support for common misconceptions and how to address them.

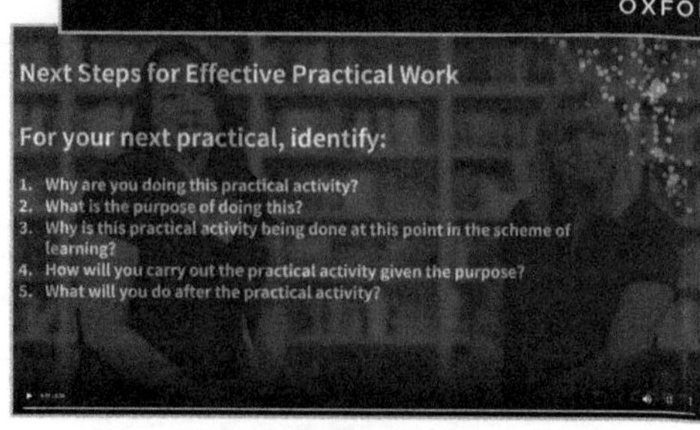

Curriculum narratives

The Curriculum narratives are overarching domain narratives, that provide a macro-level view of a learner's journey through the curriculum for each domain of science. They give a broad picture of the scope and progression of learning within the curriculum and articulate the rationale for the inclusion of content.

Working together with experienced, subject-expert teachers and reviewers, the Curriculum narratives were created using research and data mapping every part of the content to its prerequisite knowledge from the start to end of secondary education.

They have helped us ensure effective and coherent sequencing of learning for each science domain through the curriculum, informing on all resources. They also been of crucial importance in ensuring the KS3 curriculum content seamlessly follows on from primary science content and aligns with student and teacher needs at GCSE.

All the Curriculum narratives can be found in the Research hub on Kerboodle.

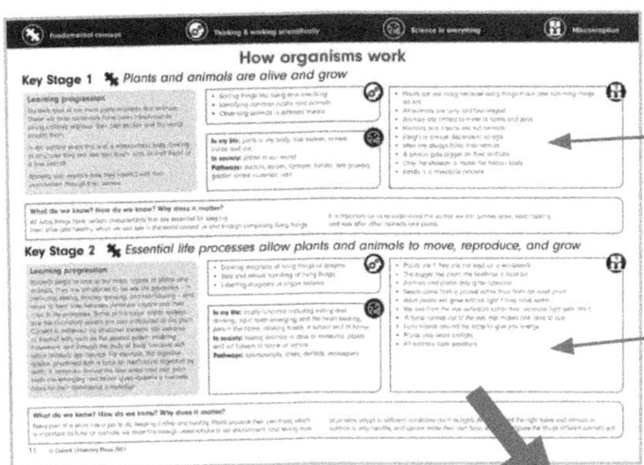

The Key Stage 1 and 2 information sets Key Stage 3 learning in the context of prior content covered and can help you to understand what knowledge and misconceptions students may bring into secondary education.

The narrative follows the same structure for each key stage.

The ***Fundamental concept*** section is a short summary of the core substantive knowledge for the domain at each key stage.

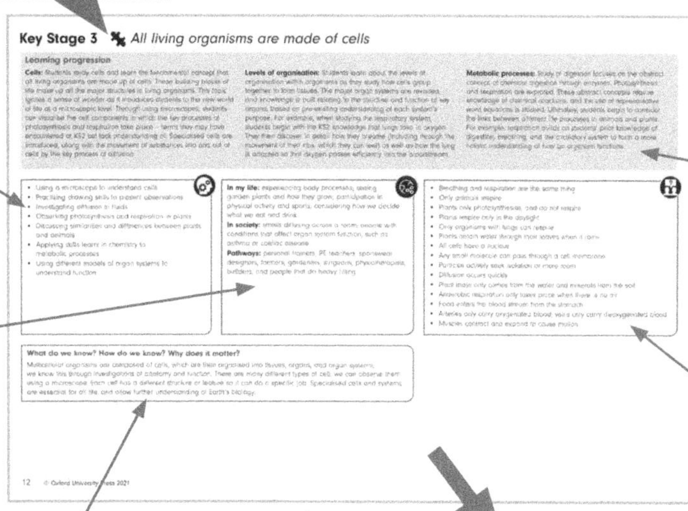

Opportunities for students to practice ***thinking and working scientifically***.

The ***Science in everything*** section gives relevancies of the domain content to learners' personal lives, wider society, and future pathways.

The ***Learning progression*** section details how the domain's substantive knowledge is introduced to learners, explored, and developed, from Key Stage 1 (where appropriate) to the end of Key Stage 4.

Some of the most common and key ***misconceptions*** learners may have about the domain's knowledge.

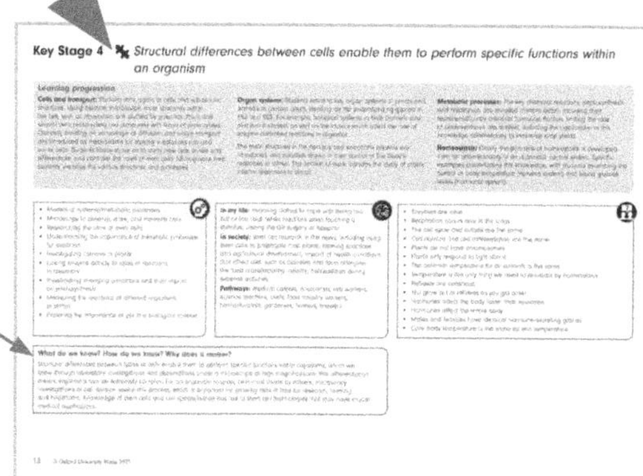

The ***What do we know? How do we know? Why does it matter?*** section addresses the rationale behind the knowledge in the domain and its importance to society and our understanding of the world.

The Key Stage 4 information sets Key Stage 3 learning in context of what learning is still to come and how students' understanding will develop.

Assessment and reporting

About the Curriculum Editor and Assessment Expert

Dr Andrew Chandler-Grevatt has a doctorate in school assessment and a passion for science teaching. Having worked as a teacher for ten years, Andy has a real understanding of teaching in the classroom. This stays at the front of his mind during all of his work in education. Alongside his research in school assessment, Andy is a teaching fellow on the PGCE course at the University of Sussex and a successful published assessment author.

Welcome from the Curriculum Editor

Welcome to your *Oxford Smart* Activate 3 Teacher Handbook. The Teacher Handbooks, together with Kerboodle, and the Student Books, provide comprehensive assessment that is coherent with the rest of the *Oxford Smart* Curriculum for Science. Our assessment is built in across the curriculum and will help your school to monitor progress against a clear framework. Throughout *Oxford Smart* Activate, formative assessment has been made easy and we have followed a set of guiding principles.

Oxford Smart Activate assessment principles

Assessment in *Oxford Smart* Activate aims to:

- inform teaching and/or learning directly (have a formative function)
- assess agreed and shared objectives
- provide opportunities for peer- and self-assessment
- provide opportunities for specific feedback to be given to individual students
- provide usable data that informs teachers of class and individual progress.

I have been working closely with our expert author teams across all components of the *Oxford Smart* Curriculum for Science, meaning you can be confident when using *Oxford Smart* Activate to monitor your students' progress.

Andy Chandler-Grevatt

Assessing the curriculum

The *Oxford Smart* Activate assessment model is based on bands: the middle band indicates that students have a secure grasp of the knowledge or skills specified in the curriculum. The band working towards 'secure' is 'developing', and the band moving past 'secure' is 'extending'.

All the lesson outcomes are differentiated. Support sheets are provided to help developing students to access the lesson activity or practical, while extension tasks and resources can be found on the Lesson presentations and Kerboodle.

The Summary Questions and End-of-Chapter Questions in the Student Book are ramped. The level of demand of each question is indicated by the number of dots shaded in by the question number.

Activate Bands	Developing	Secure	Extending
What it looks like – content	Demonstrates knowledge and understanding by recall and simple descriptions	Demonstrates secure knowledge and understanding of the learning objective	Demonstrates knowledge and understanding beyond secure
What it looks like – application	Able to apply knowledge and understanding only in highly scaffolded, familiar situations	Able to apply knowledge to familiar situations	Able to apply knowledge and understanding in unfamiliar situations

The checkpoint system

Our assessment model combines regular formative and traditional summative assessments. An additional element is regular, low-stakes quizzing aimed at interrupting the forgetting curve and helping students retain new concepts.

The formative assessment comprises the checkpoints and mini-checkpoints on Kerboodle. Regular mini-checkpoint assessment tracks progress through the chapter and there is a checkpoint assessment following each chapter. These are all automarked and will help determine if students have a secure understanding of the chapter.

Following each checkpoint or mini-checkpoint, students are offered personalised next steps. They can consolidate their knowledge where they are at a developing level and challenge themselves where they have demonstrated secure knowledge.

At the end of each year, there is a summative assessment, available as a digital or paper-based assessment, to evaluate understanding of the whole unit.

Reporting and insights

The formative assessment data will feed online reporting on Kerboodle and give insights into students' strengths and weaknesses.

The curriculum breaks each science into the Curriculum narrative domains and then splits them into related themes that run through all key stages. The themes are divided into sub-themes to enable targeted interventions.

The reporting will support you in diagnosing learner strengths and weaknesses, and to identify misconceptions so you can adapt your teaching.

Follow assessment with learning

Oxford Smart Activate has been designed with responsive teaching in mind. There are opportunities embedded throughout the course to use the formative assessment data to inform your teaching.

The data from the mini-checkpoints and checkpoints can be used to inform which follow-up activities for students to target their weaknesses and to support all students towards working at a secure level.

Extension resources are available on Kerboodle to help students at a secure level move towards extending.

The Metacognition chapter has information on building in student reflection, and this can be useful before and after an assessment.

Your Teacher Handbook

As in the previous books in the series, each unit begins with a unit opener, setting the scene for the biology, chemistry, or physics content in the book and providing lots of useful information to help you to inspire and engage your students.

Each chapter starts with a chapter opener summarising the chapter and giving useful tips to help you to deliver the content, including common misconceptions, strategies for teaching particular information, and on tackling literacy, mathematical, and working scientifically skills.

Each chapter then ends with a checkpoint lesson containing suggested activities to use as follow-up to the Checkpoint assessments to ensure that all of your students have reached a secure level.

In *Oxford Smart* Activate Book 3, there are three types of lessons, Reactivate, Apply, and Reach. The purpose of each of these was introduced on page vi.

Reactivate

These lessons are designed to reactivate students' knowledge of key concepts from *Oxford Smart* Activate Books 1 and 2. In the Student Book, this consists of a knowledge organizer and some questions. In the Teacher handbook, a few different options for activities are suggested to allow for classes who may need to spend more than one lesson on the Reactivate portion of the chapter.

The main Reactivate lesson structure revolves around a pinchpoint question. A pinchpoint is an idea or concept in science that can be challenging to learn. It is often difficult to say why these ideas are challenging to learn. Each pinchpoint intervention question focuses on a challenging idea from *Oxford Smart* Activate Books 1 or 2 that is important for the forthcoming chapter. Each question has multiple-choice answers with one correct answer and others that correspond to misconceptions, misunderstandings, and missing knowledge. There is then a follow-up activity specifically designed to address the mistake the student has made or to provide stretch for students who selected the correct answer. The pinchpoint lesson is supported by Kerboodle resources, including a lesson presentation and worksheets. The teacher and technician worksheet includes further detail on what misconception/misunderstanding each incorrect answer corresponds to and how the follow-up activity is designed to develop their understanding.

Apply

The Apply lessons take content from *Oxford Smart* Activate 1 and 2 and apply it to a new context. These contexts may involve new applications of skills.

The lessons follow the same structure as the lessons in *Oxford Smart* Activate 1 and 2 but the exposition, rather than introducing new content, may include further recap of Book 1 or 2 content, the introduction of the new context, a recap of skills to be used in the new context or modelling of those application skills.

Reach

The Reach lessons introduce new, Key Stage 4 content to bridge the gap between Key Stage 3 and 4. The structure of the lesson is the same as that in *Oxford Smart* Activate 1 and 2.

Unit opener

Overview

The Unit opener provides an overview of the unit (biology, chemistry, or physics), how it links to Key Stage 2 and 4, and ideas for how a teacher could put learning into context, as well as inspire awe and wonder with information related to the corresponding Student Book unit opener.

Introduction
An overview of the chapters in this unit.

Curriculum links
The National Curriculum subject and Working scientifically statements that link to this unit.

Learning journey
An overview of the topics covered in the unit, the Key Stage 2 topics that lead to them, and the later Key Stage 3 and Key Stage 4 topics that they build to.

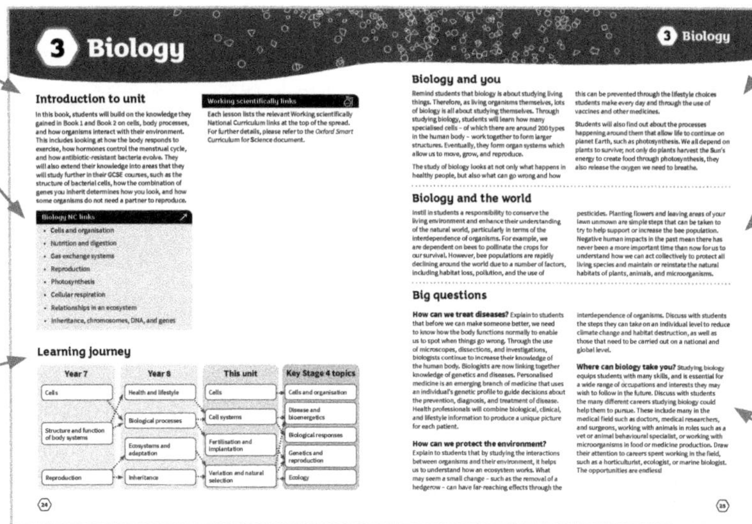

Science and you
Ideas for how to provide context for the unit content and how to make it relevant to students' lives.

Science and the world
Ideas for how to provide context for the unit content and how to make it relevant to students' lives.

Big questions
The Student Book features three fundamental questions that science aims to answer that link to the unit. This section provides further detail and guidance on leading discussions around these.

Chapter opener

Overview

The Chapter Opener provides teacher-focused preparation for the chapter content to ensure all teachers are prepared and confident to teach the content and to emphasise curriculum coherence across the course.

Core concepts
A list of the core knowledge concepts covered in this chapter.

To see how these develop throughout a students' school life, see the Curriculum narratives in the *Oxford Smart* Curriculum for Science.

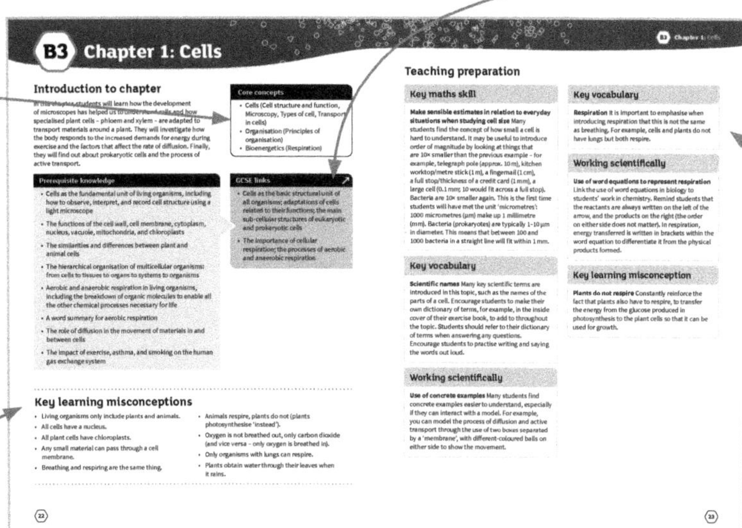

Curriculum links
The prerequisite knowledge for the chapter content (Key Stage 3 or Primary) and links to GCSE content.

Teaching preparation
Tips and key information for a teacher to consider when teaching the chapter, specifically designed to ensure that all teachers are confident in delivering the content and skills.

These will vary from chapter to chapter but may include further information on misconceptions, suggestions for tackling key skills, and useful strategies for teaching particular content.

Key learning misconceptions
A list of some of the common misconceptions that students might hold related to the chapter content to help teachers to diagnose and address them.

Further information on how these have been identified and how support is embedded in the curriculum can be found on the Kerboodle CPD and Research Hub.

Your Teacher Handbook

Reactivate lesson

Overview

Core concepts
The theme and sub-theme indicate the area of the *Oxford Smart* Curriculum for Science that this lesson focuses on. Students' understanding can be monitored with the curriculum reports. Links to the National Curriculum are also given here.

Skill icons

Pinchpoint lesson
The grey boxes in each section of the lesson plan indicate the structure of the lesson using the pinchpoint question (see page xii). This lesson is fully supported by resources on Kerboodle.

Objectives
The learning objectives being reactivated from Books 1 and 2 are listed here. Curriculum reporting on Kerboodle could be used to monitor students' prior understanding.

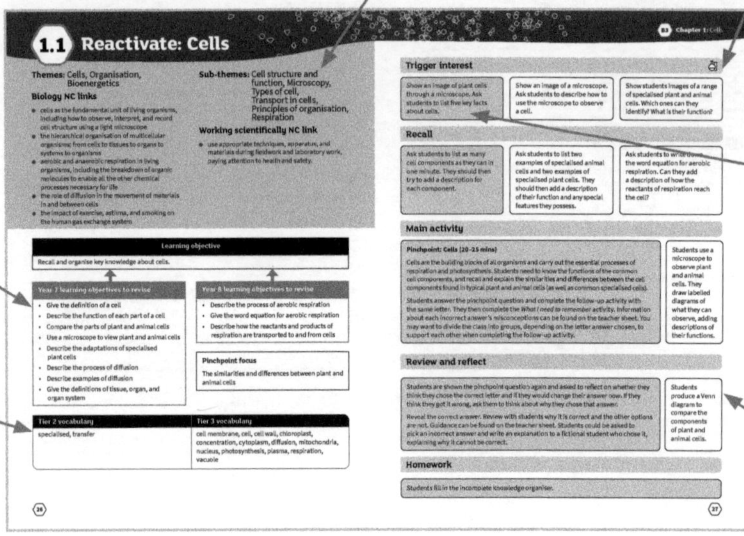

Vocabulary

Alternate activities
For each part of the lesson, there are one or two suggested alternative activities that could be used if a class needs more than one Reactivate lesson. These are not supported by Kerboodle resources.

Apply lesson

Core concepts

Skill icons

Objectives
This section includes:
- a new Apply learning objective for the lesson
- a list of the Book 1 and 2 learning objectives being applied in the lesson
- a description of the application context.

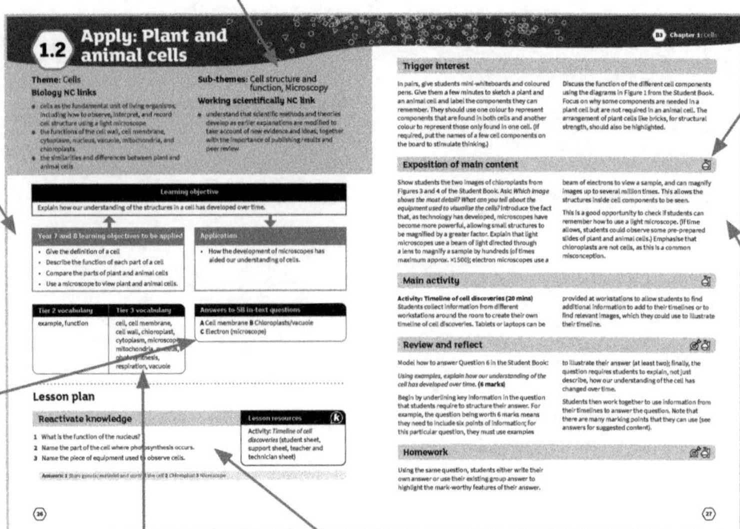

Suggested lesson plan
A suggested route through the lesson is provided, including ideas for support and homework. The structure has been carefully designed using cognitive science research and is fully supported by resources on Kerboodle.

Answers
Answers to the Student Book in-text questions can be found here. The answers to all other questions are found at the back of the Teacher Handbook and on Kerboodle.

Vocabulary

Reactivate knowledge
The retrieval questions from the Student Book are included here, along with the answers and any suggestions for discussions of key prerequisite knowledge.

Reach lesson

Objectives
This section includes:
- a new Reach learning objective for the lesson
- a list of the prerequisite Book 1 and 2 learning objectives being extended in the lesson
- how this lesson links to Key Stage 4.

Core concepts

Suggested lesson plan
A suggested route through the lesson is provided, including ideas for support and homework. The structure has been carefully designed using cognitive science research and is fully supported by resources on Kerboodle.

Vocabulary
Tier 2 vocabulary includes words that are more academic in nature and operate across different contexts. Tier 3 key words are subject-specific. Words in italics have multiple meanings and misconceptions may arise due to this.

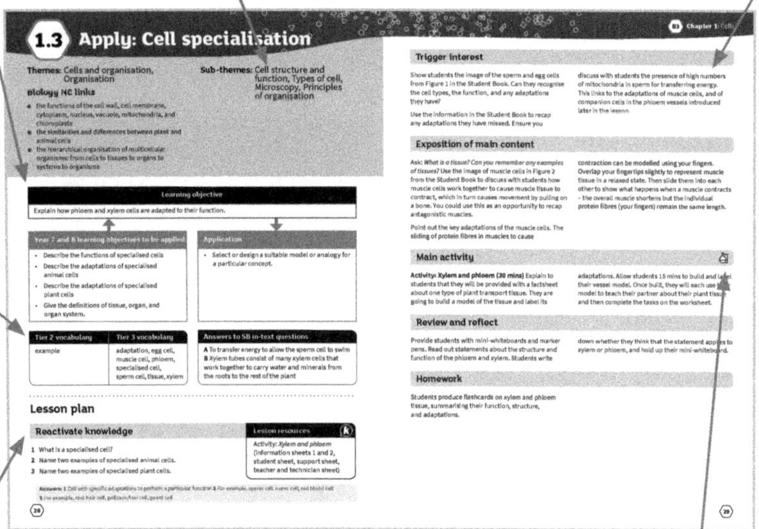
1.3 Apply: Cell specialisation

Trigger interest

Exposition of main content

Main activity

Review and reflect

Homework

Learning objective

Lesson plan

Reactivate knowledge

Reactivate knowledge

Skills
These symbols indicate where different skills have been embedded in the lesson.

Working scientifically   Literacy   Maths

These skills are ramped through *Oxford Smart* Activate and divided into the three bands. A progression grid and learning outcomes are supplied on Kerboodle.

Checkpoint lesson

Overview

The checkpoint intervention lesson is a suggested follow-up after students have completed the automarked checkpoint assessment on Kerboodle. The activities are aimed to move all students from developing to secure on the key learning objectives covered in the assessment.

Intervention activities
Guidance on possible missing information or misconceptions and an intervention activity for each core learning outcome from the checkpoint assessment with the aim to ensure that all student are at a secure level.

These activities are flexible so you can use your students' data to select the most appropriate one(s) for your class.

Checkpoint secure learning outcomes
A list of all of the secure Apply learning outcomes covered in the checkpoint assessment.

Student reflection
This section suggests how to help students to reflect on their learning over the chapter. Additional strategies to aid reflection can be found in the Metacognition chapter or in the Research hub on Kerboodle.

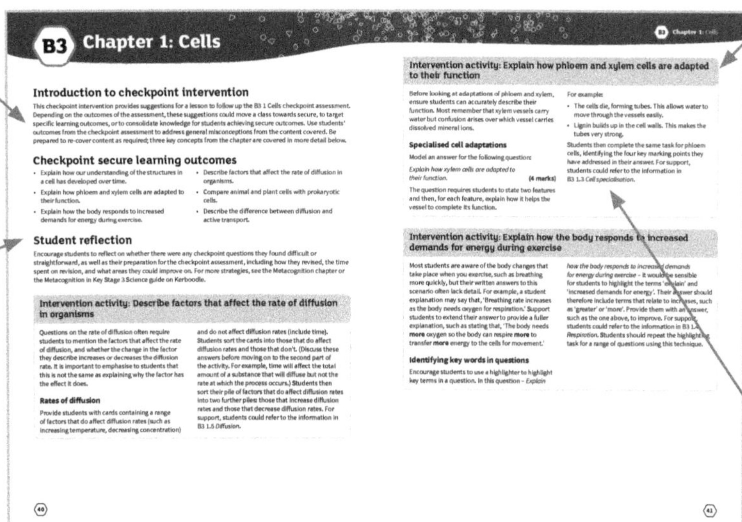
B3 Chapter 1: Cells

Introduction to checkpoint intervention

Checkpoint secure learning outcomes

Student reflection

Intervention activity: Describe factors that affect the rate of diffusion in organisms

Intervention activity: Explain how phloem and xylem cells are adapted to their function

Intervention activity: Explain how the body responds to increased demands for energy during exercise

Extending students
Further resources are available on Kerboodle for classes or students where moving from secure to extending would be appropriate.

Kerboodle

***Oxford Smart* Activate Kerboodle** provides hundreds of engaging lesson resources, as well as a comprehensive assessment package. They are regularly updated, intuitive, and customisable. Resources include lessons and associated resources, automatically marked assessments, books, and the CPD and Research hub.

You can adapt many of the resources to suit your students' needs or your own scheme of work, with all non-interactive activities available as editable Word documents. You can also upload your existing resources so that everything can be accessed from one location.

Kerboodle is online, allowing you and your students to access the course anytime and anywhere. You can set homework and assessments through the Assessment system and track progress using the Markbook.

Kerboodle Book

The *Oxford Smart* Activate Kerboodle Book provides a digital copy of the Student Book for use at the front of the classroom.

Teacher access to the Kerboodle Book is automatically available as part of the Lessons, Resources, and Assessment package. You can also purchase additional access for your students.

A set of tools is available with the Kerboodle Book so you can personalize your book and make notes.

Like all other resources offered on Kerboodle, the Kerboodle Book can also be accessed using a range of devices.

Use different tools such as sticky notes, bookmarks, and pen features to personalise each page

Navigate around the book quickly with the contents menu, key word search, or page number search.

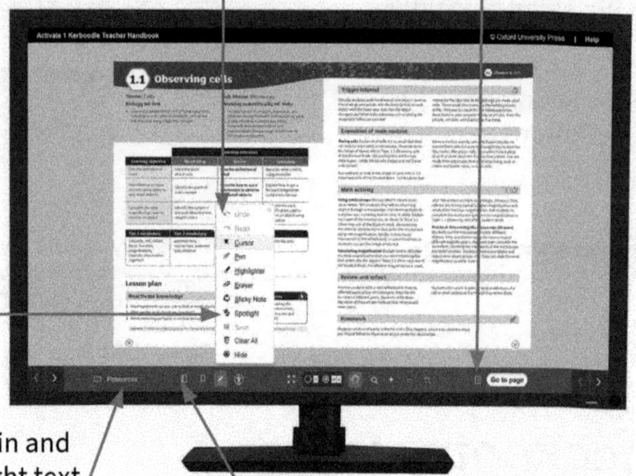

Zoom in and spotlight text

All the associated resources can be viewed on each spread.

Every teacher and student has their own digital notebook for use within their Kerboodle Book. You can choose to share your notes with students or hide them from view – all student notes are accessible to themselves only.

Lessons

Click on the 'Lessons' tab to access the *Oxford Smart* Activate Lesson Plan and Presentation (with accompanying notes).

There is a lesson plan to complement every spread in the Teacher Handbook and Student Book. Each plan is accompanied by teachers' notes on that lesson and links to the resources required for the lesson, including the Lesson presentation.

Each lesson comes with a presentation that features *Reactivate* question starters, lesson objectives, *Trigger Interest* content, activity guidance, key diagrams, extension questions, plenaries, and homework suggestions. You can further personalise these by adding in your own resources and notes.

Resources are built into each lesson plan so that associated interactive content, presentations, practical, or activity worksheets are ready to launch from one location.

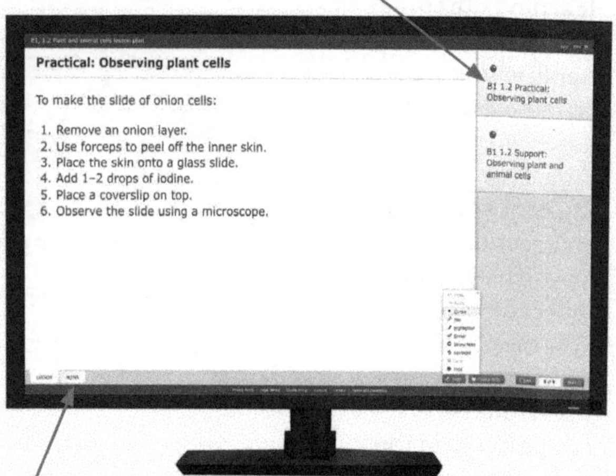

Every lesson is accompanied by teacher notes that provide additional support and extension opportunities to fully support lesson delivery.

Resources

Click on the 'Resources' tab to access the full list of *Oxford Smart* Activate lesson resources.

The resource section includes:

- **Practicals and activities** Fully-editable resources to use in class. In addition to an activity sheet and a support sheet, which are accessible by pupils, a teacher and technician sheet provides further ideas on differentiation, answers and a list of resources required by technicians.
- **Maths and Working scientifically skill sheets and interactives**
- **Videos and animations**
- **PowerPoints**
- **Extension resources**
- **Knowledge organisers**
- **Checklists**
- **Glossaries**
- **Webquests**
- **Metacognition support**

Navigation panel and search bar allow for easy navigation between resources by book, unit, and chapter.

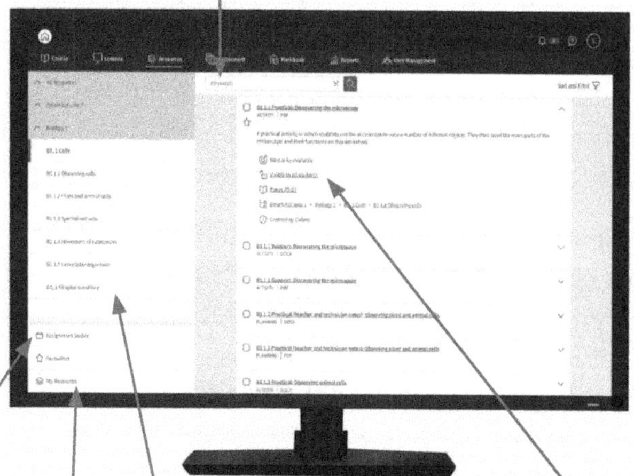

Resources can be made visible or hidden to students and a link will take you to the corresponding Student Book spread.

Set single or multiple assignments for your classes using the Assignment basket.

Existing resources and links can be uploaded onto the platform using the My Resources section.

Resources matching every lesson in the *Oxford Smart* Activate series are shown here.

Assessment and Markbook

All the assessment material in Kerboodle has been quality assured by our assessment expert. Click on the Assessment tab to find a range of materials to help you deliver a varied, motivating, and effective assessment programme.

It's easy to import class registers and create user accounts for your students. Once your classes are set up, you can assign them assessments to do at home, individually, or as a group.

A Markbook with reporting function helps you to keep track of your students' results. This includes both automarked assessments and work that needs to be marked by you.

In the Assessment tab, you can find mini-checkpoints for each lesson, checkpoints for each chapter, and end-of-unit assessments for each unit.

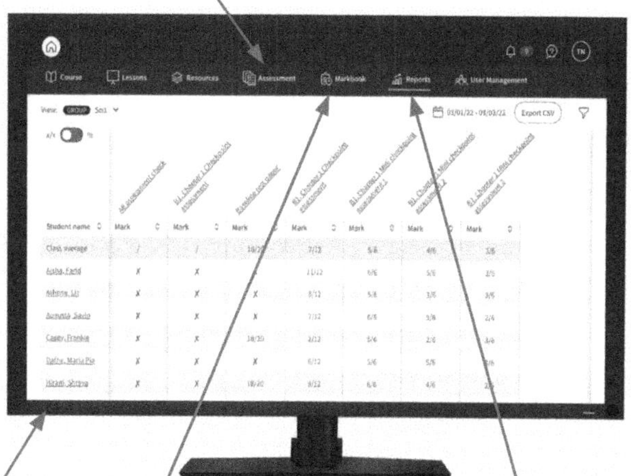

Different follow-up tasks are available to students after they complete the checkpoints and mini-checkpoints, depending on how well they did.

A Markbook and reporting function helps you track your students' progress.

Reports can be generated using data from the checkpoints and mini-checkpoints.

Preparing for an assessment

What does 'learning' really mean?

For students to 'learn' a piece of information, they need to be able to store it within their long-term memory and recall it when needed. One way we can enhance this process for students is by understanding how our memory works so that we can guide students through using more effective learning strategies.

The Multi-store model of memory outlines the main components of memory and how information can be transferred between different memory stores. Our sensory memory is where information that is first detected through our senses is stored. By paying attention to the information, it is then transferred into our short-term memory store. If we need to use that information we then use our working memory to complete a task.

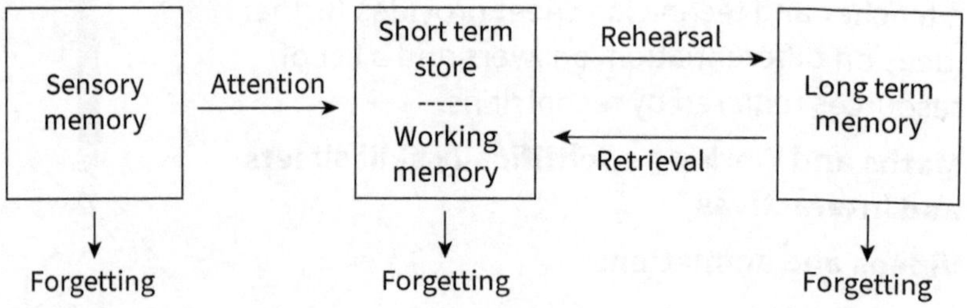

▲**Figure 1** A diagram showing the Multi-store model of memory

When we conduct a new or complicated task, we can sometimes find that our working memory becomes quickly overloaded. This hinders the chances of that information being transferred to long-term memory and therefore the information can be easily forgotten. Information can be transferred to our long term memory through rehearsal and transferred back to our working memory by retrieval. The more often we rehearse and retrieve information the stronger the memory becomes.

In the next section there are several effective revision strategies outlined. The one thing that they all have in common is that they are grounded in retrieval. It is important to get students to understand that the more often that they recall information, however difficult it may be at first, the stronger the learning of that information becomes and the more they will be able to recall over time.

Effective revision strategies

In 2013, John Dunlosky published *Strengthening the Student Toolbox*[1], within which he explored study strategies to boost learning. Dunlosky's research focuses on how self-regulated learning can impact student attainment. He identifies the following strategies as highly effective for learning:

- **Practice testing:** Where students are repeatedly carrying out tests, quizzes, and practice exam questions to recall and apply information.
- **Distributed practice:** Where students spread out their revision topics over a period of time rather than spending lots of time revising one thing. This helps them recall the content over time.
- **Interleaved practice:** Where students swap around the order of topics when they are revising so that they are covering each topic little and often.
- **Elaborative interrogation:** Where students delve deeper into content and start to make connections between topics or attempt to explain why a process takes place.

1 J. Dunlosky (2013), *Strengthening the Student Toolbox*, American Educator

Dunlosky identifies the common methods of rereading, summarising, and highlighting text as some of the least effective ways students can revise – even if these are their typical go-to methods of revision.

For students to understand their own learning processes, they need to understand **why** strategies are useful as well as how to use them. One way to introduce the above to students is to model their use in revision lessons. Deliberately planning a series of revision sessions that follow these strategies, combined with sharing the explanation of **why** they are so effective, will give students greater metacognitive awareness and improve their self-regulation of their learning.

The Leitner method

One of the strategies outlined in the student book is the Leitner method for using flash cards.

Here is one way you can introduce this method with students following the seven-step approach:

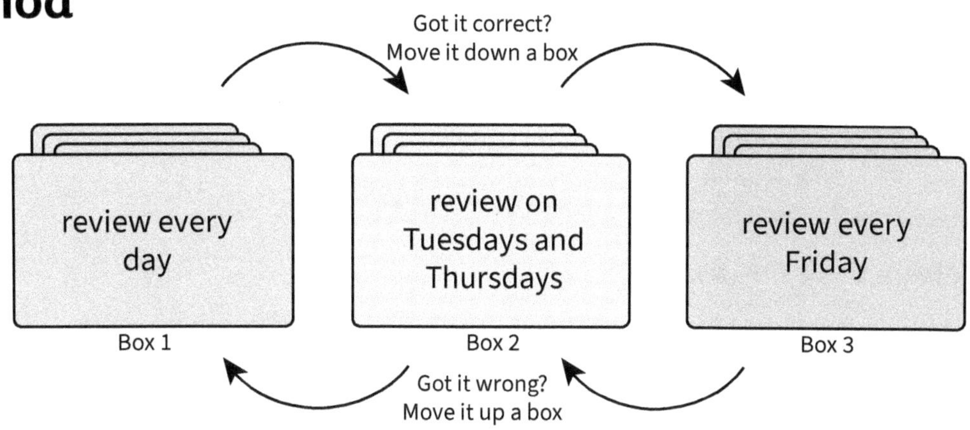

▲ **Figure 2** The Leitner method

1. Activate prior knowledge	The teacher reviews key content relating to the topic using low stakes quizzes, topic maps, and some independent reading.
2. Explicit strategy instruction	The teacher explores what makes a good flash card. The teacher then introduces the Leitner method to the students. Each step of the method is explained by linking to the theory of why recall activities like this are so important for learning. Doing this helps the students understand why the Leitner method is such an effective strategy.
3. Modelling of learned strategy	The teacher then models the Leitner method to the students, reiterating the previous explanation of the steps.
4. Memorisation of strategy	The students are given time to note down the strategy and then the teacher circulates the room and questions students to see if they understand how to use the strategy and why it is effective.
5. Guided practice	The class is then given a selection of premade flash cards and practice the Leitner method using these.
6. Independent practice	Next the students are given the opportunity to create a few of their own flash cards to then practice the Leitner method with. This could then lead into a homework task where students need to create a set of flash cards for an upcoming assessment and show their use of the Leitner method.
7. Structured reflection	At the end of the lesson, the group reflects on how effective the Leitner method was in comparison to simple quizzing. They can also reflect on how it can be used for other topics or subjects, feeding it into a revision plan for future assessments.

Tackling assessments like a pro

Exam technique

Have you ever found that students are able to recall subject knowledge well, but seem to fall at the final hurdle of being able to apply it to a particular question? Answering exam questions is a skill and needs to be explicitly taught, just like we would teach many of the other skills in science.

When a student first comes across an exam question, it can be overwhelming and it can easily overload their working memory. Teaching them a specific technique to follow can be really beneficial. There are many acronyms and techniques that can be used but the BUG technique is a particularly easy one to follow.

BUG

Box the command word
Underline any key information
Go over the question and your answer

▲**Figure 1** The BUG technique

The seven-step approach from the Education Endowment Foundation can be used to introduce and embed the BUG technique:

1. Activate prior knowledge
2. Explicit strategy instruction
3. Modelling of learned strategy
4. Memorisation of strategy
5. Guided practice
6. Independent practice
7. Structured reflection

A key aspect of this approach is modelling. Modelling metacognitively focuses on demonstrating how an expert learner would approach the task by verbalising their inner thought processes. An example of this when demonstrating the BUG technique would be to do a walking talking mock. This is where the teacher completes an exam paper 'live' in front of the students, using the BUG technique where appropriate. Students are then able to mirror the strategy and thought processes themselves when completing questions.

After the exam – Evaluating performance

The period of time after an exam is a good opportunity to support students in developing their metacognitive knowledge to become more self-regulated. To do this, students should be given time to reflect on and evaluate their own performance.

An exam wrapper is a document which scaffolds the post-exam review process by using questions and prompts to guide students through identifying subject content and skills that they need to review. An exam wrapper also helps students to reflect on their preparation by asking students when and how they revised. It also provides students with an opportunity to plan for how they will move their own learning forward by asking them to write 'next steps'.

Total mark: ____________ Percentage: ____________

Topic/Skill	Score

Which topic(s) have you highlighted above as being your key area(s) of focus moving forward? Are these old or new topics?

____

____

Are there any procedural skills you need to focus on?

____

____

How did you prepare for this assessment?

Strategy	How much time?	Strategy	How much time?
Self-quizzing KO		Practice questions	
Revision roulette		Reading notes	

Next steps: Before your next assessment, you need to review the topics you have picked out. You need to make sure you are using effective revision techniques and are addressing any procedural skills you may need to work on. Plan and write down your next steps.

▲ **Figure 2** An example of an exam wrapper

After the exam – feedback

Here are some examples of activities that can get students actively engaging with feedback.

Find and fix

This is where students are actively seeking to find and correct mistakes in model answers before turning their attention to their own work. This can be done by first collating common mistakes into one document when marking the assessment. Students are then asked to then find and correct mistakes, before reviewing their own answers to make improvements.

Find and fix the mistakes

Uncertainty is: 1120 – 960 = 160	An electrical component that has resistance increase as more current flows through it is known as a LDR.	As the temperature of air inside the canister increases, the pressure decreases, so the canister will be crushed.

▲ **Figure 3** An example of a Find and fix

Three questions

This is where you write three targeted questions for each student which delve into their misconceptions and mistakes. Asking students to focus on answering these rather than reviewing the whole test paper focuses their attention on the areas that they most need to improve upon.

3 Working scientifically

Introduction to the unit

In this unit, students will look again at the main stages of a scientific investigation, starting with identifying variables, producing a plan to create valid data, and performing a risk assessment. They will practise calculating averages as well as presenting data on a range of charts, including bar charts, pie charts, and histograms. Finally, they will analyse data to draw conclusions, carry out evaluations to identify any sources of errors, and suggest improvements to their methods.

In addition to this unit, working scientifically skills have been embedded within lessons. Each lesson spread includes the relevant Working scientifically National Curriculum links.

Working scientifically NC links

- scientific attitudes
- experimental skills and investigations
- analysis and evaluation
- measurement

Learning journey

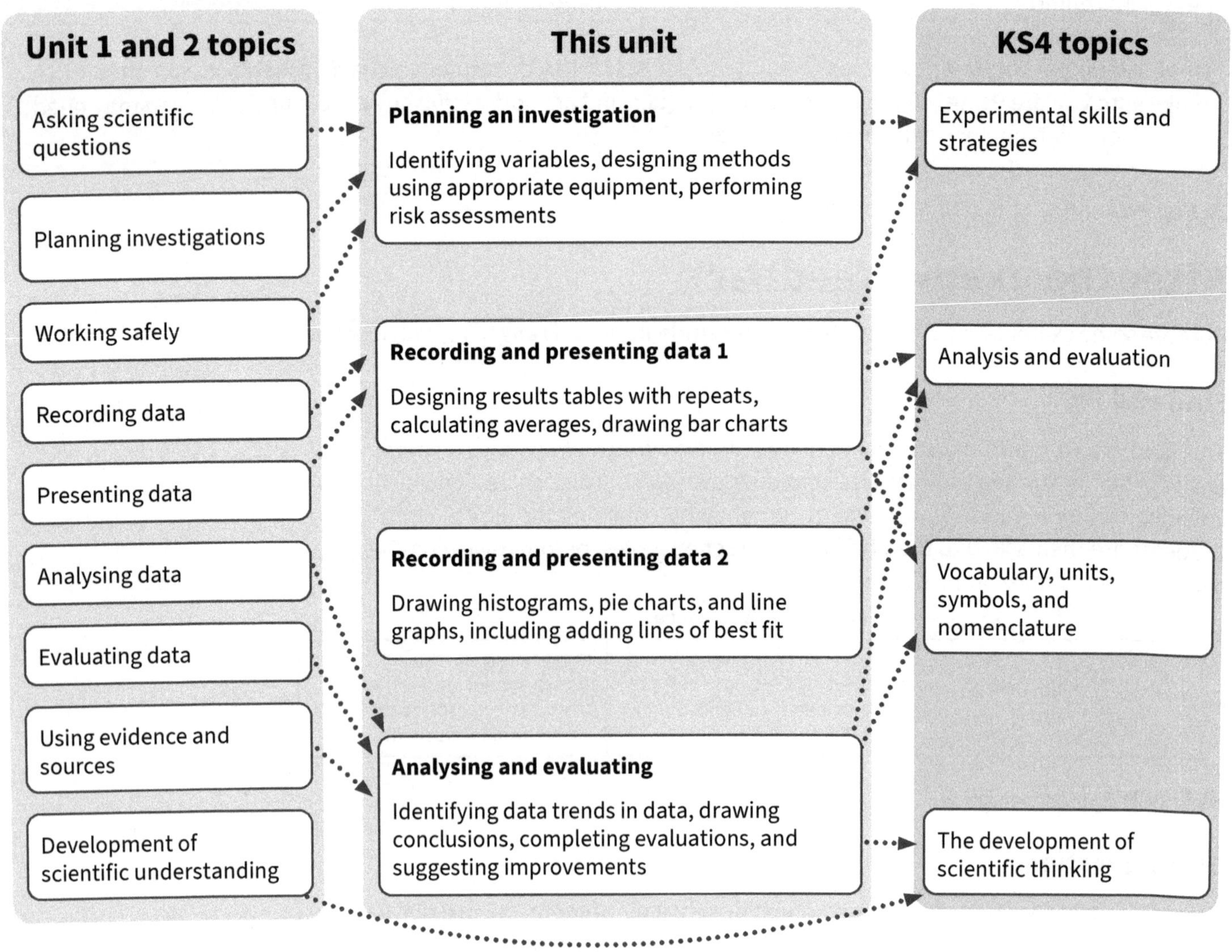

Working scientifically and you

Discuss the fact that humans are naturally scientific. We are born with inquisitive minds, constantly asking questions to make sense of the world in which we live. By carrying out scientific investigations based on the observations they make, people can start to find reasons for why things occur or predict what will happen in a situation.

Understanding the scientific process and how to interpret data is an important skill for everyday life. It allows people to make informed choices and decisions about things like their health and lifestyle.

Discuss that by knowing how to analyse and evaluate problems and solutions, it can make people more sustainable, as they can find ways to repair, reuse, and recycle things.

Working scientifically and the world

It is important to stress that there are still many things that we cannot explain fully and new problems arise, which scientists endeavour to solve, so scientific research is constantly occurring. This happens in the laboratory, in the field, and even in Space. Due to advances in technology and improvements in communications, many research projects are now global, as illustrated by the fight against COVID-19. This is increasingly important with the shared stand against climate change and its wide-ranging effects.

Scientists also have a key role in exposing 'fake news', making sure that the population is well-informed using authentic scientific information. They also provide data to governments to enable them to make decisions based on evidence.

Big questions

How do scientists keep safe? Discuss with students the role of risk assessments when carrying out an investigation in order to protect both the scientist and others working around them. These identify the hazards and steps that can be taken to minimise the risk of these hazards causing harm. It is important that students are aware that most investigations involve a degree of risk. However, many safety measures are employed to minimise this.

How can you trust a scientific claim? Explain to students that one of the key skills of a scientist is to think critically. When looking at a claim, they should think about what evidence there is for a claim, what the basis is for the claim, who wrote it, and why. Only when they know the answers to these questions can they make a reasoned decision on whether to believe a piece of information, or not.

Why do scientists change their minds about why things happen? Explain that scientists are constantly questioning the work of scientists who have come before them. Technology is constantly developing, for example, the development of electron microscopes from the single glass magnifying lens, or the ability to sequence DNA. New technology combined with carrying out different experiments and new ways of thinking can all provide new evidence. This can result in scientists having to amend a previous idea or change their thinking entirely.

3.1 Planning an investigation

Working scientifically NC links

- ask questions and develop a line of enquiry based on observations of the real world, alongside prior knowledge and experience
- make predictions using scientific knowledge and understanding
- select, plan, and carry out the most appropriate types of scientific enquiries to test predictions, including identifying independent, dependent, and control variables, where appropriate

Learning objective

Recall and organise key knowledge about planning an investigation.

Year 7 learning objectives to revise and apply

- Develop an idea into a question that can be investigated.
- Identify independent, dependent, and control variables.
- Make a scientific prediction.
- Recognise commonly used hazard symbols.
- Identify possible hazards in practical work.
- Understand the purpose of a risk assessment.
- Write a plan for a scientific investigation.

Year 8 learning objectives to revise and apply

- Write a hypothesis for a scientific investigation.
- Describe the difference between precise and accurate data.
- Write a risk assessment for a scientific investigation.

Tier 2 vocabulary	Tier 3 vocabulary
accurate, control measure, data, dependent variable, hazard, hypothesis, independent variable, observations, plan, precise, prediction, range, risk, variable	control variable, random error, systematic error

Lesson plan

Reactivate knowledge

Lesson resources

Activity: *Investigating candle burn time - planning* (student sheet, support sheet, teacher and technician sheet)

1 What is an independent variable?
2 What is a dependent variable?
3 What is meant by a hazard?
4 What is meant by precise data?

Answers: 1 The variable you change in an investigation **2** The variable that is changed by changing the independent variable **3** Something which may cause injury **4** Data which is similar when repeating a measurement

Trigger interest

Show students the list of hazard symbols, along with their hazard names. Ask students to match the hazard to the symbol.

Discuss with students why hazard symbols are important and a range of jobs in which people may encounter hazard symbols.

Exposition of main content

Introduce students to the investigation they are going to complete: Which factors affect how long a candle burns? You may wish to complete a brief demonstration to show students, in general terms, how they will complete their investigation. You may also wish to provide a range of equipment students could use to complete their investigation.

Explain to students that they will take four lessons to complete the overall investigation, focusing on:

- planning (this lesson)
- collecting data (lesson 2)
- analysing results (lesson 3)
- evaluating the investigation (lesson 4).

Working in pairs, ask students to recall the main information they should include in a plan. Share this as a class, asking each pair to add a key feature. This is a good opportunity to check students' understanding of the different types of variable.

Main activity

Activity: Investigating candle burn time - planning (35 mins) Working in small groups, students work through the investigation planning activity sheet to produce a detailed plan for their choice of investigation.

Review and reflect

Describe a number of key terms used in a plan, such as a risk assessment, dependent variable, or hypothesis.

Students write down the term being described on their mini-whiteboard and hold it up.

Homework

Students complete their copy of the knowledge organiser.

3.2 Recording and presenting data 1

Working scientifically NC links

- use appropriate techniques, apparatus, and materials during fieldwork and laboratory work, paying attention to health and safety
- make and record observations and measurements using a range of methods for different investigations; and evaluate the reliability of methods and suggest possible improvements
- present observations and data using appropriate methods, including tables and graphs

Learning objective

Recall and organise key knowledge about carrying out an investigation.

Year 7 learning objectives to revise and apply

- Design a results table.
- Present results in a correctly designed results table.
- Calculate a mean from repeat measurements.

Year 8 learning objective to revise and apply

- Calculate the mean, median, and mode of a set of data.

Tier 2 vocabulary	Tier 3 vocabulary
anomalous result, bar chart, data, line graph, mean, median, mode, pie chart	*categorical data*, *continuous data*, *discrete data*, histogram

Lesson plan

Reactivate knowledge

1 Which variable goes in the first column of a results table?

2 What is an anomalous result?

3 What are the three types of average?

4 What is categorical data?

5 Which variable goes on the y-axis?

Lesson resources

Practical: *Investigating candle burn time - experiment* (student sheet, support sheet, teacher and technician sheet)

Answers: 1 Independent variable **2** A result that is very different to the others when repeating measurement **3** Mean, mode, and median **4** Data where the value is a word **5** Dependent variable

Trigger interest

Show students the results table of how the mass of a sycamore seed affects how long it takes to fall. The table contains a number of mistakes, such as the independent variable should be in the left-hand column and the mass requires a unit. Ask students to spot the mistakes and share their findings. As a class, rewrite the table on the board in the correct form.

Exposition of main content

Remind students of the practical investigation started in the previous lesson: Which factors affect how long a candle burns? Remind them of their progress through the unit of work, and that the focus of this lesson is on collecting and recording data:

- planning (lesson 1)
- collecting data (this lesson)
- analysing results (lesson 3)
- evaluating the investigation (lesson 4)

Ask students to re-read their method and risk assessment from the previous lesson. Remind students of the importance of using a range of (five) values of the independent variable, and of using (three) repeat readings in an investigation.

Ask students to 'mock up' a design for their results table using mini-whiteboards. Share some good examples with the class. This will also allow you to identify students who may require additional support in creating their results table.

Main activity

Practical: Investigating candle burn time - experiment (35–40 mins)
Working in small groups, students follow the instructions on the practical sheet. This asks students to design their results table, including selecting values to be used for the independent variable, and then complete their practical investigation. There are also a series of follow-up questions, including calculating a mean candle burn time value for each value of their independent variable.

Review and reflect

Call out a number of key terms linked to collecting and recording data. Students should consider how the term was applied to their investigation and note this on their mini-whiteboards. For example:

Number of values for the independent variable (5)

Number of repeats (3)

Order of the independent variable in results table (smallest to largest)

Where to write units in a results table (heading line, after the quantity name)

How do you identify an anomaly? (a result that is very different to the other repeats)

Homework

Students complete their copy of the knowledge organiser.

3.3 Recording and presenting data 2

Working scientifically NC links

- present observations and data using appropriate methods, including tables and graphs
- interpret observations and data, including identifying patterns and using observations, measurements, and data to draw conclusions
- present reasoned explanations, including explaining data in relation to predictions and hypotheses

Learning objective

Recall and organise key knowledge about presenting data.

Year 7 learning objectives to revise and apply

- Classify data as continuous, discrete, or categorical.
- Present data as a chart or graph.
- Identify patterns or trends in data using a graph or chart.
- Interpret data to draw conclusions.

Year 8 learning objectives to revise and apply

- Select the appropriate graph to display data.
- Identify linear and directly proportional relationships.

Tier 2 vocabulary	Tier 3 vocabulary
anomalous result, bar chart, dependent variable, independent variable, line graph, pie chart	*categorical data*, *continuous data*, directly proportional, *discrete data*, histogram, line of best fit, linear relationship

Lesson plan

Reactivate knowledge

1 What data is represented using a histogram?
2 Which variable is plotted on the x-axis of a graph?
3 How do you calculate the angle of a pie chart segment?
4 What is a line of best fit?
5 How do you calculate frequency density?
6 What should you include on a graph axis?

Answers: 1 Grouped continuous data **2** Independent variable **3** Divide segment value by total value, then multiply by 360 **4** A trendline that goes through, or close to, as many points as possible **5** Frequency divided by class width **6** Name of variable, units, appropriate scale

Lesson resources

Activity: *Investigating candle burn time - analysis* (student sheet, support sheet, teacher and technician sheet)

Trigger interest

Show students the graph in the Lesson presentation of students' attendance against mean GCSE science grade. Ask students what conclusions they can draw from this data. Most students should identify that there is a positive relationship between the two variables. However, challenge students to discuss the flattening out of the curve. This is also a good opportunity to discuss the difference between correlation and causation.

Exposition of main content

Remind students of the practical investigation started in the previous lessons: Which factors affect how long a candle burns? Remind them of their progress through the unit of work and that the focus of this lesson is on analysing results:

- planning (lesson 1)
- collecting data (lesson 2)
- analysing results (this lesson)
- evaluating the investigation (lesson 4)

Ask students to look again at their results from the last lesson. Ask: *What initial conclusions can you draw from looking at your data table?* Students should discuss this in their groups.

Their next task is to decide which type of graph is the most appropriate to use to plot their data. Show students a range of different data using the Lesson presentation. Discuss which plot would be the most appropriate for each data. Remind students that line graphs are only plotted when both the independent and dependent variables are continuous.

Main activity

Investigating candle burn time-analysis (35 mins)
Continuing to work in their practical groups, students follow the tasks on the activity worksheet. These ask students to select the most appropriate chart to present their data, and then complete an analysis of their data. Students complete this exercise by relating the data collected to their hypothesis.

Review and reflect

Show students the graph in the Lesson presentation of how the mass of magnesium added to 50 cm^3 hydrochloric acid affects the temperature rise during a reaction. Students evaluate the positioning of the line of best fit on the graph. To prompt thinking, ask students to think about what would happen to the temperature if no magnesium were added.

Homework

Students complete their copy of the knowledge organiser.

3.4 Analysing and evaluating

Working scientifically NC links

- interpret observations and data, including identifying patterns and using observations, measurements, and data to draw conclusions
- present reasoned explanations, including explaining data in relation to predictions and hypotheses
- evaluate data, showing awareness of potential sources of random and systematic error

Learning objective
Recall and organise key knowledge about evaluating an investigation.

Year 7 learning objectives to revise and apply

- Identify patterns or trends in data using a graph or chart.
- Interpret data to draw conclusions.
- Describe the stages in evaluating data.
- Suggest ways to improve a practical investigation.

Year 8 learning objectives to revise and apply

- Calculate the mean, mode, and median of a set of data.
- Describe how to improve confidence in a conclusion.

Tier 2 vocabulary	Tier 3 vocabulary
anomalous result, bar chart, conclusion, evaluation, improvement, line graph, range, spread, strength, trend, weakness	peer review, random error, repeatable, secondary data, systematic error

Lesson plan

Reactivate knowledge

1 What type of graph shows a linear relationship?

2 What is a conclusion?

3 What causes a random error?

4 What is secondary data?

Answers: 1 A straight-line graph **2** A description of what you found out **3** An unexpected change in experimental conditions **4** Data that has been collected by someone else

Lesson resources

Activity: *Investigating candle burn time - evaluation* (student sheet, support sheet, teacher and technician sheet)

Trigger interest

Show students the results table in the Lesson presentation of how the concentration of an acid affects the time taken to neutralise an alkali. This table contains a number of mistakes. Ask students to spot the mistakes and redraw the table correctly. Some are more straightforward to spot, such as the units not being stated in the column heading. Others require students to work through the calculations of the mean.

Exposition of main content

Remind students of the practical investigation started in the previous lessons: Which factors affect how long a candle burns? Remind them of their progress through the unit of work and that the focus of this lesson is on evaluating their data and their overall investigation:

- planning (lesson 1)
- collecting data (lesson 2)
- analysing results (lesson 3)
- evaluating the investigation (this lesson)

Ask students to look back over their practical method, results tables, graph, and conclusions from the last three lessons. Remind students that scientists must have confidence in their data, so that there is real evidence from which to form a conclusion. In this lesson, students will be evaluating their data and their investigation as a whole.

Ask students to share what they can remember about the steps needed in the evaluation of scientific data. Remind students that experiments are repeated to smooth out small differences in data. Only larger differences are classified as anomalous results; these should be ignored when calculating a mean. A large spread in data or the presence of many anomalies undermines confidence in a conclusion.

Recap the difference between random and systematic errors. Discuss general techniques for improving a method, such as the use of different equipment or taking measurements for a larger range of the independent variable.

Main activity

Investigating candle burn time-evaluation (35–40 mins)
Working in their practical groups, students follow the instructions on the activity sheet. This takes students through the process of evaluating their data and their investigation.

They then evaluate data from a second investigation on how the number of turns on a wind-up toy car affected how far the car travelled. If time allows, this would be a quick investigation students could complete themselves to generate raw data, which they could then evaluate.

Review and reflect

Show students the results table again from the start of the lesson.

Explain that the students who completed this investigation concluded that, 'All acids will neutralise alkalis more quickly if they are more concentrated'.

Ask students to discuss the validity of this conclusion. Have they enough data to make generalisations? If time allows, students could be asked to write a more appropriate conclusion for this data.

Homework

Students complete their copy of the knowledge organiser.

3 Biology

Introduction to unit

In this book, students will build on the knowledge they gained in Book 1 and Book 2 on cells, body processes, and how organisms interact with their environment. This includes looking at how the body responds to exercise, how hormones control the menstrual cycle, and how antibiotic-resistant bacteria evolve. They will also extend their knowledge into areas that they will study further in their GCSE courses, such as the structure of bacterial cells, how the combination of genes you inherit determines how you look, and how some organisms do not need a partner to reproduce.

Biology NC links

- Cells and organisation
- Nutrition and digestion
- Gas exchange systems
- Reproduction
- Photosynthesis
- Cellular respiration
- Relationships in an ecosystem
- Inheritance, chromosomes, DNA, and genes

Working scientifically links

Each lesson lists the relevant Working scientifically National Curriculum links at the top of the spread. For further details, please refer to the *Oxford Smart Curriculum for Science* document.

Learning journey

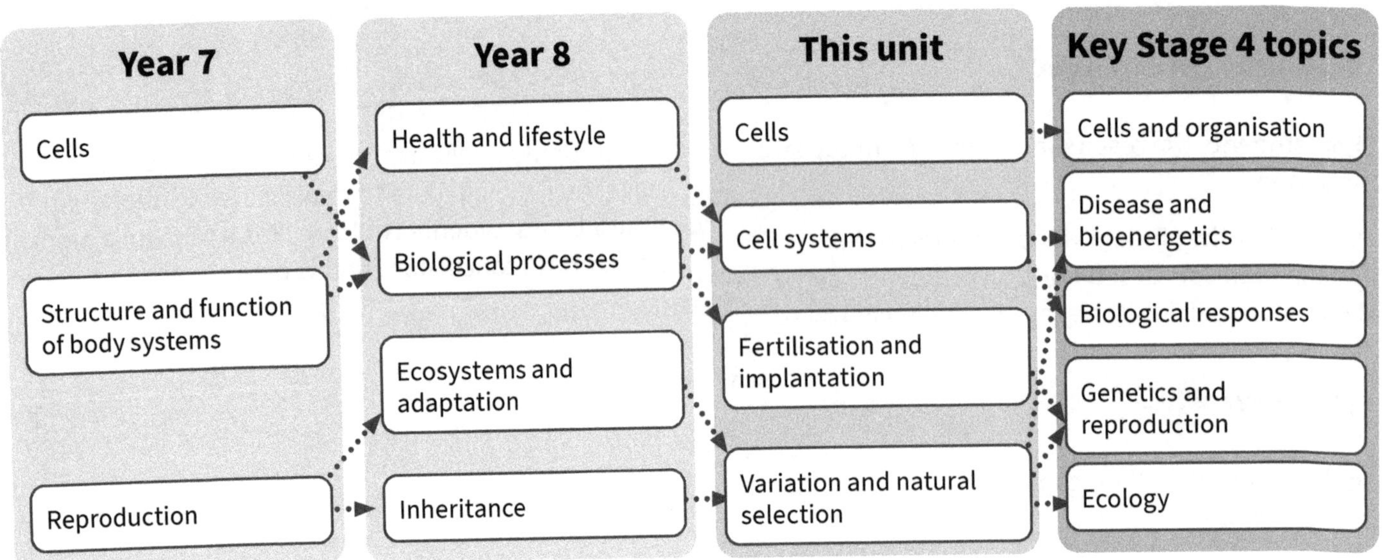

Biology and you

Remind students that biology is about studying living things. Therefore, as living organisms themselves, lots of biology is all about studying themselves. Through studying biology, students will learn how many specialised cells – of which there are around 200 types in the human body – work together to form larger structures. Eventually, they form organ systems which allow us to move, grow, and reproduce.

The study of biology looks at not only what happens in healthy people, but also what can go wrong and how this can be prevented through the lifestyle choices students make every day and through the use of vaccines and other medicines.

Students will also find out about the processes happening around them that allow life to continue on planet Earth, such as photosynthesis. We all depend on plants to survive; not only do plants harvest the Sun's energy to create food through photosynthesis, they also release the oxygen we need to breathe.

Biology and the world

Instil in students a responsibility to conserve the living environment and enhance their understanding of the natural world, particularly in terms of the interdependence of organisms. For example, we are dependent on bees to pollinate the crops for our survival. However, bee populations are rapidly declining around the world due to a number of factors, including habitat loss, pollution, and the use of pesticides. Planting flowers and leaving areas of your lawn unmown are simple steps that can be taken to try to help support or increase the bee population. Negative human impacts in the past mean there has never been a more important time than now for us to understand how we can act collectively to protect all living species and maintain or reinstate the natural habitats of plants, animals, and microorganisms.

Big questions

How can we treat diseases? Explain to students that before we can make someone better, we need to know how the body functions normally to enable us to spot when things go wrong. Through the use of microscopes, dissections, and investigations, biologists continue to increase their knowledge of the human body. Biologists are now linking together knowledge of genetics and diseases. Personalised medicine is an emerging branch of medicine that uses an individual's genetic profile to guide decisions about the prevention, diagnosis, and treatment of disease. Health professionals will combine biological, clinical, and lifestyle information to produce a unique picture for each patient.

How can we protect the environment? Explain to students that by studying the interactions between organisms and their environment, it helps us to understand how an ecosystem works. What may seem a small change – such as the removal of a hedgerow – can have far-reaching effects through the interdependence of organisms. Discuss with students the steps they can take on an individual level to reduce climate change and habitat destruction, as well as those that need to be carried out on a national and global level.

Where can biology take you? Studying biology equips students with many skills, and is essential for a wide range of occupations and interests they may wish to follow in the future. Discuss with students the many different careers studying biology could help them to pursue. These include many in the medical field such as doctors, medical researchers, and surgeons, working with animals in roles such as a vet or animal behavioural specialist, or working with microorganisms in food or medicine production. Draw their attention to careers spent working in the field, such as a horticulturist, ecologist, or marine biologist. The opportunities are endless!

B3 Chapter 1: Cells

Introduction to chapter

In this chapter, students will learn how the development of microscopes has helped us to understand cells and how specialised plant cells – phloem and xylem – are adapted to transport materials around a plant. They will investigate how the body responds to the increased demands for energy during exercise and the factors that affect the rate of diffusion. Finally, they will find out about prokaryotic cells and the process of active transport.

Core concepts

- Plant and animal cells
- Specialised cells
- Aerobic respiration
- Diffusion and osmosis

Prerequisite knowledge

- Cells as the fundamental unit of living organisms, including how to observe, interpret, and record cell structure using a light microscope
- The functions of the cell wall, cell membrane, cytoplasm, nucleus, vacuole, mitochondria, and chloroplasts
- The similarities and differences between plant and animal cells
- The hierarchical organisation of multicellular organisms: from cells to tissues to organs to systems to organisms
- Aerobic and anaerobic respiration in living organisms, including the breakdown of organic molecules to enable all the other chemical processes necessary for life
- A word summary for aerobic respiration
- The role of diffusion in the movement of materials in and between cells
- The impact of exercise, asthma, and smoking on the human gas exchange system

GCSE links

- Cells as the basic structural unit of all organisms; adaptations of cells related to their functions; the main sub-cellular structures of eukaryotic and prokaryotic cells
- The importance of cellular respiration; the processes of aerobic and anaerobic respiration

Key learning misconceptions

- Living organisms only include plants and animals.
- All cells have a nucleus.
- All plant cells have chloroplasts.
- Any small material can pass through a cell membrane.
- Breathing and respiring are the same thing.
- Animals respire, plants do not (plants photosynthesise 'instead').
- Oxygen is not breathed out, only carbon dioxide (and vice versa – only oxygen is breathed in).
- Only organisms with lungs can respire.
- Plants obtain water through their leaves when it rains.

Teaching preparation

Key maths skill

Make sensible estimates in relation to everyday situations when studying cell size Many students find the concept of how small a cell is hard to understand. It may be useful to introduce order of magnitude by looking at things that are 10× smaller than the previous example – for example, telegraph pole (approx. 10 m), kitchen worktop/metre stick (1 m), a fingernail (1 cm), a full stop/thickness of a credit card (1 mm), a large cell (0.1 mm; 10 would fit across a full stop). Bacteria are 10× smaller again. This is the first time students will have met the unit 'micrometres': 1000 micrometres (μm) make up 1 millimetre (mm). Bacteria (prokaryotes) are typically 1–10 μm in diameter. This means that between 100 and 1000 bacteria in a straight line will fit within 1 mm.

Key vocabulary

Scientific names Many key scientific terms are introduced in this topic, such as the names of the parts of a cell. Encourage students to make their own dictionary of terms, for example, in the inside cover of their exercise book, to add to throughout the topic. Students should refer to their dictionary of terms when answering any questions. Encourage students to practise writing and saying the words out loud.

Working scientifically

Use of concrete examples Many students find concrete examples easier to understand, especially if they can interact with a model. For example, you can model the process of diffusion and active transport through the use of two boxes separated by a 'membrane', with different-coloured balls on either side to show the movement.

Key vocabulary

Respiration It is important to emphasise when introducing respiration that this is not the same as breathing. For example, cells and plants do not have lungs but both respire.

Working scientifically

Use of word equations to represent respiration Link the use of word equations in biology to students' work in chemistry. Remind students that the reactants are always written on the left of the arrow, and the products on the right (the order on either side does not matter). In respiration, energy transferred is written in brackets within the word equation to differentiate it from the physical products formed.

Key learning misconception

Plants do not respire Constantly reinforce the fact that plants also have to respire, to transfer the energy from the glucose produced in photosynthesis to the plant cells so that it can be used for growth.

1.1 Reactivate: Cells

Themes: Bioenergetics, Cells, Organisation

Biology NC links

- cells as the fundamental unit of living organisms, including how to observe, interpret, and record cell structure using a light microscope
- the hierarchical organisation of multicellular organisms: from cells to tissues to organs to systems to organisms
- aerobic and anaerobic respiration in living organisms, including the breakdown of organic molecules to enable all the other chemical processes necessary for life
- the role of diffusion in the movement of materials in and between cells
- the impact of exercise, asthma, and smoking on the human gas exchange system

Sub-themes: Cell structure and function, Microscopy, Principles of organisation, Respiration, Transport in cells, Types of cell

Working scientifically NC link

- use appropriate techniques, apparatus, and materials during fieldwork and laboratory work, paying attention to health and safety.

Learning objective
Recall and organise key knowledge about cells.

Year 7 learning objectives to revise

- Give the definition of a cell
- Describe the function of each part of a cell
- Compare the parts of plant and animal cells
- Use a microscope to view plant and animal cells
- Describe the adaptations of specialised plant cells
- Describe the process of diffusion
- Describe examples of diffusion
- Give the definitions of tissue, organ, and organ system

Year 8 learning objectives to revise

- Describe the process of aerobic respiration
- Give the word equation for aerobic respiration
- Describe how the reactants and products of respiration are transported to and from cells

Pinchpoint focus

The similarities and differences between plant and animal cells

Tier 2 vocabulary	Tier 3 vocabulary
cell, concentration, specialised, transfer	cell membrane, cell wall, chloroplast, cytoplasm, *diffusion*, mitochondria, nucleus, photosynthesis, plasma, respiration, vacuole

Trigger interest

Show an image of plant cells through a microscope. Ask students to list five key facts about cells.

Show an image of a microscope. Ask students to describe how to use the microscope to observe a cell.

Show students images of a range of specialised plant and animal cells. Which ones can they identify? What is their function?

Recall

Ask students to list as many cell components as they can in one minute. They should then try to add a description for each component.

Ask students to list two examples of specialised animal cells and two examples of specialised plant cells. They should then add a description of their function and any special features they possess.

Ask students to write down the word equation for aerobic respiration. Can they add a description of how the reactants of respiration reach the cell?

Main activity

Pinchpoint: Cells (20–25 mins)

Cells are the building blocks of all organisms and carry out the essential processes of respiration and photosynthesis. Students need to know the functions of the common cell components, and recall and explain the similarities and differences between the cell components found in typical plant and animal cells (as well as common specialised cells).

Students answer the pinchpoint question and complete the follow-up activity with the same letter. They then complete the *What I need to remember* activity. Information about each incorrect answer's misconceptions can be found on the teacher sheet. You may want to divide the class into groups, depending on the letter answer chosen, to support each other when completing the follow-up activity.

Students use a microscope to observe plant and animal cells. They draw labelled diagrams of what they can observe, adding descriptions of their functions.

Review and reflect

Students are shown the pinchpoint question again and asked to reflect on whether they think they chose the correct letter and if they would change their answer now. If they think they got it wrong, ask them to think about why they chose that answer.

Reveal the correct answer. Review with students why it is correct and the other options are not. Guidance can be found on the teacher sheet. Students could be asked to pick an incorrect answer and write an explanation to a fictional student who chose it, explaining why it cannot be correct.

Students produce a Venn diagram to compare the components of plant and animal cells.

Homework

Students fill in the incomplete knowledge organiser.

1.2 Apply: Plant and animal cells

Theme: Cells

Sub-themes: Cell structure and function, Microscopy

Biology NC links

- cells as the fundamental unit of living organisms, including how to observe, interpret, and record cell structure using a light microscope
- the functions of the cell wall, cell membrane, cytoplasm, nucleus, vacuole, mitochondria, and chloroplasts
- the similarities and differences between plant and animal cells

Working scientifically NC link

- understand that scientific methods and theories develop as earlier explanations are modified to take account of new evidence and ideas, together with the importance of publishing results and peer review

Learning objective

Explain how our understanding of the structures in a cell has developed over time.

Year 7 and 8 learning objectives to be applied

- Give the definition of a cell
- Describe the function of each part of a cell
- Compare the parts of plant and animal cells
- Use a microscope to view plant and animal cells.

Application

- How the development of microscopes has aided our understanding of cells.

Tier 2 vocabulary	Tier 3 vocabulary
cell, example, function	cell membrane, cell wall, chloroplast, cytoplasm, microscope, mitochondria, nucleus, photosynthesis, respiration, vacuole

Answers to SB in-text questions

A Cell membrane **B** Chloroplasts/vacuole
C Electron (microscope)

Lesson plan

Reactivate knowledge

1 What is the function of the nucleus?
2 Name the part of the cell where photosynthesis occurs.
3 Name the piece of equipment used to observe cells.

Answers: 1 Store genetic material and control the cell **2** Chloroplast **3** Microscope

Lesson resources

Activity: *Timeline of cell discoveries* (student sheet, support sheet, teacher and technician sheet)

Trigger interest

In pairs, give students mini-whiteboards and coloured pens. Give them a few minutes to sketch a plant and an animal cell and label the components they can remember. They should use one colour to represent components that are found in both cells and another colour to represent those only found in one cell. (If required, put the names of a few cell components on the board to stimulate thinking.)

Discuss the function of the different cell components using the diagrams in Figure 1 from the Student Book. Focus on why some components are needed in a plant cell but are not required in an animal cell. The arrangement of plant cells like bricks, for structural strength, should also be highlighted.

Exposition of main content

Show students the two images of chloroplasts from Figures 3 and 4 of the Student Book. Ask: *Which image shows the most detail? What can you tell about the equipment used to visualise the cells?* Introduce the fact that, as technology has developed, microscopes have become more powerful, allowing small structures to be magnified by a greater factor. Explain that light microscopes use a beam of light directed through a lens to magnify a sample by hundreds (of times maximum approx. ×1500); electron microscopes use a beam of electrons to view a sample, and can magnify images up to several million times. This allows the structures inside cell components to be seen.

This is a good opportunity to check if students can remember how to use a light microscope. (If time allows, students could observe some pre-prepared slides of plant and animal cells.) Emphasise that chloroplasts are not cells, as this is a common misconception.

Main activity

Activity: Timeline of cell discoveries (20 mins)
Students collect information from different workstations around the room to create their own timeline of cell discoveries. Tablets or laptops can be provided at workstations to allow students to find additional information to add to their timelines or to find relevant images, which they could use to illustrate their timeline.

Review and reflect

Model how to answer Question 6 in the Student Book:

Using examples, explain how our understanding of the cell has developed over time. **(6 marks)**

Begin by underlining key information in the question that students require to structure their answer. For example, the question being worth 6 marks means they need to include six points of information; for this particular question, they must use examples to illustrate their answer (at least two); finally, the question requires students to explain, not just describe, how our understanding of the cell has changed over time.

Students then work together to use information from their timelines to answer the question. Note that there are many marking points that they can use (see answers for suggested content).

Homework

Using the same question, students either write their own answer or use their existing group answer to highlight the mark-worthy features of their answer.

1.3 Apply: Cell specialisation

Themes: Cells, Organisation

Sub-themes: Cell structure and function, Microscopy, Principles of organisation, Types of cell

Biology NC links

- the functions of the cell wall, cell membrane, cytoplasm, nucleus, vacuole, mitochondria, and chloroplasts
- the similarities and differences between plant and animal cells
- the hierarchical organisation of multicellular organisms: from cells to tissues to organs to systems to organisms

Learning objective

Explain how phloem and xylem cells are adapted to their function.

Year 7 and 8 learning objectives to be applied

- Describe the functions of specialised cells
- Describe the adaptations of specialised animal cells
- Describe the adaptations of specialised plant cells
- Give the definitions of tissue, organ, and organ system.

Application

- Building a model of xylem and phloem.

Tier 2 vocabulary	Tier 3 vocabulary
example	adaptation, egg cell, muscle cell, phloem, specialised cell, sperm cell, *tissue*, xylem

Answers to SB in-text questions

A To transfer energy to allow the sperm cell to swim
B Xylem tubes consist of many xylem cells that work together to carry water and minerals from the roots to the rest of the plant

Lesson plan

Reactivate knowledge

1 What is a specialised cell?
2 Name two examples of specialised animal cells.
3 Name two examples of specialised plant cells.

Lesson resources

Activity: *Xylem and phloem* (information sheets × 2, student sheet, support sheet, teacher and technician sheet)

Answers: 1 Cell with specific adaptations to perform a particular function **2** For example, sperm cell, nerve cell, red blood cell **3** For example, root hair cell, palisade/leaf cell, guard cell

Trigger interest

Show students the image of the sperm and egg cells from Figure 1 in the Student Book. Can they recognise the cell types, the function, and any adaptations they have?

Use the information in the Student Book to recap any adaptations they have missed. Ensure you discuss with students the presence of high numbers of mitochondria in sperm for transferring energy. This links to the adaptations of muscle cells, and of companion cells in the phloem vessels introduced later in the lesson.

Exposition of main content

Ask: *What is a tissue? Can you remember any examples of tissues?* Use the image of muscle cells in Figure 2 from the Student Book to discuss with students how muscle cells work together to cause muscle tissue to contract, which in turn causes movement by pulling on a bone. You could use this as an opportunity to recap antagonistic muscles.

Point out the key adaptations of the muscle cells. The sliding of protein fibres in muscles to cause contraction can be modelled using your fingers. Overlap your fingertips slightly to represent muscle tissue in a relaxed state. Then slide them into each other to show what happens when a muscle contracts – the overall muscle shortens but the individual protein fibres (your fingers) remain the same length.

Main activity

Activity: Xylem and phloem (30 mins) Explain to students that they will be provided with a factsheet about one type of plant transport tissue. They are going to build a model of the tissue and label its adaptations. Allow students 15 mins to build and label their vessel model. Once built, they will each use their model to teach their partner about their plant tissue and then complete the tasks on the worksheet.

Review and reflect

Provide students with mini-whiteboards and marker pens. Read out statements about the structure and function of the phloem and xylem. Students write down whether they think that the statement applies to xylem or phloem, and hold up their mini-whiteboard.

Homework

Students produce flashcards on xylem and phloem tissue, summarising their function, structure, and adaptations.

1.4 Apply: Respiration

Theme: Bioenergetics

Sub-theme: Respiration

Biology NC links

- aerobic and anaerobic respiration in living organisms, including the breakdown of organic molecules to enable all the other chemical processes necessary for life
- a word summary for aerobic respiration

Working scientifically NC links

- use appropriate techniques, apparatus, and materials during fieldwork and laboratory work, paying attention to health and safety
- make and record observations and measurements using a range of methods for different investigations
- present observations and data using appropriate methods, including tables and graphs
- interpret observations and data, including identifying patterns and using observations, measurements, and data to draw conclusions

Learning objective

Explain how the body responds to increased demands for energy during exercise.

Year 7 and 8 learning objectives to be applied

- Describe the process of aerobic respiration
- Give the word equation for aerobic respiration
- Describe how the reactants and products of respiration are transported to and from cells.

Application

- Investigating the effects of exercise on breathing rate.

Tier 2 vocabulary	Tier 3 vocabulary
calculate, function, rate	aerobic respiration, alveoli, haemoglobin, mitochondria, plasma

Answers to SB in-text questions

A Glucose and oxygen **B** Mitochondrion
C Heart rate

Lesson plan

Reactivate knowledge

1 Where does respiration take place?
2 Name the products of aerobic respiration.
3 Name the type of respiration that occurs without oxygen.

Answers: 1 Mitochondria **2** Water and carbon dioxide **3** Anaerobic respiration

Lesson resources

Practical: *The effect of exercise on breathing* (student sheet, support sheet, teacher and technician sheet)

Trigger interest

Write the terms from the word equation for aerobic respiration on the board in the wrong order. Ask students to rearrange them to give the correct equation. Encourage students to annotate their equation to explain how each of the reactants enters the body and travels to the cell.

After checking students have formed the correct word equation, take each reactant in turn to describe its journey into the body and into the cell, using the flowchart in Figure 1 in the Student Book as a summary. Then revisit the mechanism by which the products are removed from the body.

This activity also provides opportunities to recap the changes that take place in the body when you breathe, adaptations of the alveoli and red blood cells, and the purpose of digestion.

Exposition of main content

In pairs, ask students to make a list of the changes that take place in the body when they exercise. Why do they think this happens?

Share students' answers. All ideas should be discussed, such as 'sweating to cool you down'. However, focus on changes to breathing rate and heart rate, and the role they play in taking in oxygen and transporting oxygen and glucose to the cells for respiration. Extend students' understanding by explaining that glycogen in the muscle cells is broken down, releasing glucose which is carried in the plasma to the cells.

Main activity

Practical: The effect of exercise on breathing (30 mins) Students begin by calculating their average resting breathing rate. They then investigate how their breathing rate is affected by exercise, by calculating their breathing rate at one-minute intervals after exercise. They present their data graphically and answer the questions that follow.

Review and reflect

Show students the graph in Figure 3 from the Student Book and ask them in pairs to identify two trends from the graph. Can they explain these? Students should then share their ideas with another pair to try to improve their answers. These answers should then be shared with the class to form a model explanation.

Homework

Students create a graphical organiser that illustrates how the body responds to the increased demands for energy during exercise. For support, provide students with a list of things their organiser should contain, such as the word equation for respiration, a description of what happens to heart rate, and a description of what happens to breathing rate.

1.5 Apply: Diffusion

Theme: Cells

Sub-theme: Transport in cells

Biology NC link

- the role of diffusion in the movement of materials in and between cells

Working scientifically NC links

- select, plan, and carry out the most appropriate types of scientific enquiries to test predictions, including identifying independent, dependent, and control variables
- use appropriate techniques, apparatus, and materials during fieldwork and laboratory work, paying attention to health and safety
- make and record observations and measurements using a range of methods for different investigations

Learning objective

Describe factors that affect the rate of diffusion in organisms.

Year 7 and 8 learning objectives to be applied

- Describe the process of diffusion
- Describe examples of diffusion.

Application

- Factors that affect the rate of diffusion.

Tier 2 vocabulary	Tier 3 vocabulary
concentration, randomly, rate, surface area	cell membrane, concentration gradient, *diffusion*, osmosis, partially permeable membrane

Answers to SB in-text questions

A Initially, the blue particles are concentrated in one corner of the box, as the particles move randomly they gradually start to mix with the red particles and spread out from the corner until they are eventually spread evenly throughout the area

B Increase the concentration gradient, increase the temperature, increase the surface area

Lesson plan

Reactivate knowledge

1. What is the function of the cell membrane?
2. Name some substances that diffuse into cells.
3. Name some substances that diffuse out of cells.

Lesson resources

Practical: *Investigating the rate of diffusion* (student sheet, support sheet, teacher and technician sheet)

Answers: 1 Control which substances can pass in and out of the cell **2** For example, oxygen, glucose **3** For example, carbon dioxide, water

Trigger interest

Model diffusion for students by adding two drops of blue food dye and two drops of red food dye to a test tube of water. Ask: *What do you think is going to happen? Why?*

Talk through what happened using the diagrams in Figure 1 of the Student Book. Explain that, in this scenario, diffusion is complete when the liquid in the test tube is a uniform purple colour. Ensure students are using the terms diffusion (sometimes confused with dissolving), concentration, and random particle movement correctly before moving on.

Exposition of main content

Ask students to recall some examples of where diffusion occurs in the body and its purpose. Give prompts, if necessary, such as respiration and digestion.

Using the diagrams in Figures 3 and 4 of the Student Book, explain how concentration gradient, temperature, and surface area can increase the rate of diffusion.

Then introduce the special type of diffusion – osmosis – which refers specifically to the movement of water molecules across a partially permeable membrane. This can be modelled practically using a colander containing rice (you could dye this blue) to represent water molecules, and peas to represent sugar molecules, or two different-sized beads. You then shake the colander which represents the partially permeable membrane to show only water molecules can pass through.

Main activity

Practical: Investigating the rate of diffusion (30 mins) Students design their own investigation to study the effects of temperature on the rate of diffusion based on the demonstration at the start of the lesson. Students need to design their own method, carry out their investigation, and analyse their results. They then answer questions based on their investigation.

Review and reflect

Provide students with mini-whiteboards and marker pens. Read out a series of statements, such as increasing the temperature of the solution. Students write down 'increase' or 'decrease' to describe what they think will happen to the rate of diffusion and hold up their answer.

Homework

Students answer the following exam question:

Explain how the alveoli in the lungs are adapted for diffusion. **(6 marks)**

Supply students with an annotated diagram and tips to help them structure their answer.

1.6 Reach: Prokaryotic cells

Theme: Cells

Sub-themes: Cell structure and function, Microscopy, Types of cell

Biology NC links

- cells as the fundamental unit of living organisms, including how to observe, interpret, and record cell structure using a light microscope (KS3)
- the functions of the cell wall, cell membrane, cytoplasm, nucleus, vacuole, mitochondria, and chloroplasts (KS3)
- the similarities and differences between plant and animal cells (KS3)
- cells as the basic structural unit of all organisms; adaptations of cells related to their functions; the main sub-cellular structures of eukaryotic and prokaryotic cells (KS4)

Working scientifically NC links

- use appropriate techniques, apparatus, and materials during fieldwork and laboratory work, paying attention to health and safety
- make and record observations and measurements using a range of methods for different investigations

Learning objective
Compare animal and plant cells with prokaryotic cells.

Year 7 and 8 learning objectives to be built on

- Describe the function of each part of a cell
- Compare the parts of plant and animal cells
- Use a microscope to view plant and animal cells.

Key Stage 4 link

- The structure of eukaryotic and prokaryotic cells

Tier 2 vocabulary	Tier 3 vocabulary
bacteria, compare, difference, magnification, similarity	adaptation, eukaryote, flagellum, pili, plasmid, prokaryote, slime capsule

Answers to SB in-text questions

A An organism that is made up of just one cell
B Cell membrane, cell wall, cytoplasm, and floating genetic material

Lesson plan

Reactivate knowledge

1. What are the four cell components that are found in both plant and animal cells?
2. What are the three cell components that are only found in plant cells?
3. What is the function of mitochondria?

Answers: 1 Nucleus, cell membrane, cytoplasm, mitochondria **2** Cell wall, vacuole, chloroplast **3** Respiration/to transfer energy to the cell

Lesson resources

Practical: *Observing prokaryotic cells* (student sheet, support sheet, teacher and technician sheet)

Trigger interest

Show students the image of bacteria from Figure 1 of the Student Book. Ask: *Do you know what the image is showing? Are they unicellular or multicellular organisms? Where do you think they may be found?*

Explain to students that they are looking at bacteria that are found in their intestine. They are unicellular, but some species of bacteria (like these) remain attached together, forming chains. Explain that different bacterial species can be different shapes.

Remind students that bacteria can only be observed using a microscope. Bacteria are approximately 1 µm in size (1 µm is 0.001 mm) and hundreds of thousands of them would fit on a full stop. Show the image of the agar plate from Figure 2 of the Student Book, or a real plate. Explain that the dots you can see are not individual bacteria – each dot is a bacterial colony made up of millions of bacteria.

Exposition of main content

Explain to students that bacteria are prokaryotes. Each is a single prokaryotic cell. Show students the image of a prokaryotic cell from Figure 3 of the Student Book. Ask: *Can you recognise which main feature is missing?* Hopefully, students will spot it is lacking a nucleus. (It is also missing any membrane-bound organelles, such as mitochondria and chloroplasts.) Explain that the genetic material in a prokaryotic cell is found free, floating in the cytoplasm.

Discuss the remaining similarities and differences between eukaryotic and prokaryotic cells.

Use Figure 4 of the Student Book to introduce the other adaptations that bacterial cells may have: flagella, slime capsule, pili, and plasmids. Ensure students are aware that not all bacteria cells have these features.

Main activity

Practical: Observing prokaryotic cells (30 mins)
Students observe bacteria used in yoghurt production under the microscope. They then answer questions to compare the structures found in prokaryotic and eukaryotic cells.

Review and reflect

Provide students with three different-coloured cards. For example, green to represent plant cells, red to represent animal cells, and blue to represent prokaryotic cells. Name or describe a part of a cell, such as a cell wall made of cellulose. Students hold up the card(s) representing the type(s) of cell that contain these components.

Homework

Ask students to construct a Venn diagram to compare the similarities and differences between eukaryotic and prokaryotic cells. To extend more able students, ask them to construct a three-way Venn diagram containing eukaryotic animal cells, eukaryotic plant cells, and prokaryotic cells.

1.7 Reach: Active transport

Theme: Cells

Sub-theme: Transport in cells

Biology NC links

- the role of diffusion in the movement of materials in and between cells (KS3)
- cells as the basic structural unit of all organisms; adaptations of cells related to their functions; the main sub-cellular structures of eukaryotic and prokaryotic cells (KS4)

Learning objective
Describe the difference between diffusion and active transport.

Year 7 and 8 learning objectives to be built on
• Describe the process of diffusion • Describe examples of diffusion.

Key Stage 4 link
• Transport in cells – diffusion and active transport

Tier 2 vocabulary	Tier 3 vocabulary
concentration	active transport, concentration gradient, *diffusion*, osmosis

Answers to SB in-text questions
A In diffusion, particles move from high concentration to low concentration; in active transport, particles move from low concentration to high concentration **B** The concentration of minerals in the water surrounding the roots is lower than in the plant **C** Digestion

Lesson plan

Reactivate knowledge

1 What is diffusion?
2 What are some examples of diffusion in animals?
3 What factors affect the rate of diffusion?

Lesson resources
Activity: *Active transport* (student sheet, support sheet, teacher and technician sheet)

Answers: 1 Movement of particles from a high to a low concentration **2** For example, oxygen into red blood cells, carbon dioxide into the plasma **3** Diffusion distance, concentration gradient, surface area, temperature

Trigger interest

Show students the images of the road signs in the Lesson presentation. Ask: *Which is the steeper? If you were sitting at the top of the hill on a bike and you freewheeled, in which direction would you travel? Do you need to input any energy? What biological process do you think this example models?*

Students should be able to work out that the 14% incline is the steeper gradient, and they will move from left to right without doing work. The hill gradient models diffusion.

Exposition of main content

Ask students to think how they could move in the opposite direction (up the hill). They should realise that they would need to do work to cycle up the hill. Explain that this is a model for another transport process in organisms – active transport. This is when something is moved against its concentration gradient: the process requires an input of energy. Illustrate this idea using Figure 1 in the Student Book.

Discuss the similarities and differences between active transport, diffusion, and osmosis. This is summarised in Table 1 in the Student Book.

You may wish to use one example from the Student Book to illustrate active transport, such as the uptake of mineral ions by the root hair cells in plants, but students will be researching examples as part of the activity.

Main activity

Activity: Active transport (30 mins) This activity is split into three tasks. In the first task, students write a description of active transport and complete a diagram to illustrate the process. They then sort statements about active transport, diffusion, and osmosis to compare these processes using a Venn diagram. Finally, they carry out research to find some examples of where animals and plants use active transport.

Review and reflect

Students play 'word splat'. Write the words 'diffusion', 'active transport', and 'osmosis' on three separate cards. Read out a series of statements about these processes or examples, such as 'substances moved against a concentration gradient'. In pairs, students 'splat' the card that correctly matches the statement. The one who 'splats' it first is awarded the point.

Homework

Students produce a graphic organiser to summarise the three different processes by which substances move into and out of cells: active transport, diffusion, and osmosis. The organiser should include examples of each process in both plants and animals.

Introduction to checkpoint intervention

This checkpoint intervention provides suggestions for a lesson to follow up the B3 1 *Cells* checkpoint assessment. Depending on the outcomes of the assessment, these suggestions could move a class towards secure, to target specific learning outcomes, or to consolidate knowledge for students achieving secure outcomes. Use students' outcomes from the checkpoint assessment to address general misconceptions from the content covered. Be prepared to re-cover content as required; three key concepts from the chapter are covered in more detail below.

Checkpoint secure learning outcomes

- Explain how our understanding of the structures in a cell has developed over time.
- Explain how phloem and xylem cells are adapted to their function.
- Explain how the body responds to increased demands for energy during exercise.
- Describe factors that affect the rate of diffusion in organisms.
- Compare animal and plant cells with prokaryotic cells.
- Describe the difference between diffusion and active transport.

Student reflection

Encourage students to reflect on whether there were any checkpoint questions they found difficult or straightforward, as well as their preparation for the checkpoint assessment, including how they revised, the time spent on revision, and what areas they could improve on. For more strategies, see the Metacognition chapter or the Metacognition in Key Stage 3 Science guide on Kerboodle.

Intervention activity: Describe factors that affect the rate of diffusion in organisms

Questions on the rate of diffusion often require students to mention the factors that affect the rate of diffusion, and whether the change in the factor they describe increases or decreases the diffusion rate. It is important to emphasise to students that this is not the same as explaining why the factor has the effect it does.

Rates of diffusion

Provide students with cards containing a range of factors that do affect diffusion rates (such as increasing temperature, decreasing concentration) and do not affect diffusion rates (include time). Students sort the cards into those that do affect diffusion rates and those that don't. (Discuss these answers before moving on to the second part of the activity. For example, time will affect the total amount of a substance that will diffuse but not the rate at which the process occurs.) Students then sort their pile of factors that do affect diffusion rates into two further piles: those that increase diffusion rates and those that decrease diffusion rates. For support, students could refer to the information in B3 1.5 *Diffusion*.

Intervention activity: Explain how phloem and xylem cells are adapted to their function

Before looking at adaptations of phloem and xylem, ensure students can accurately describe their function. Most remember that xylem vessels carry water but confusion arises over which vessel carries dissolved mineral ions.

Specialised cell adaptations

Model an answer for the following question:

Explain how xylem cells are adapted to their function. **(4 marks)**

The question requires students to state two features and then, for each feature, explain how it helps the vessel to complete its function.

For example:

- The cells die, forming tubes. This allows water to move through the vessels easily.
- Lignin builds up in the cell walls. This makes the tubes very strong.

Students then complete the same task for phloem cells, identifying the four key marking points they have addressed in their answer. For support, students could refer to the information in B3 1.3 *Cell specialisation*.

Intervention activity: Explain how the body responds to increased demands for energy during exercise

Most students are aware of the body changes that take place when you exercise, such as breathing more quickly, but their written answers to this scenario often lack detail. For example, a student explanation may say that, 'Breathing rate increases as the body needs oxygen for respiration.' Support students to extend their answer to provide a fuller explanation, such as stating that, 'The body needs **more** oxygen so the body can respire **more** to transfer **more** energy to the cells for movement.'

Identifying key words in questions

Encourage students to use a highlighter to highlight key terms in a question. In this question – *Explain how the body responds to increased demands for energy during exercise* – it would be sensible for students to highlight the terms 'explain' and 'increased demands for energy'. Their answer should therefore include terms that relate to increases, such as 'greater' or 'more'. Provide them with an answer, such as the one above, to improve. For support, students could refer to the information in B3 1.4 *Respiration*. Students should repeat the highlighting task for a range of questions using this technique.

B3 Chapter 2: Cell systems

Introduction to chapter

In this chapter, students will revisit the hierarchical nature of the levels of organisation within an organism. They will look at the role of enzymes in the digestive system and how they can be used commercially in washing detergents. They will discover the main features of exchange surfaces and their function in multicellular organisms, focusing on the alveoli in the lungs, villi in the intestines, and the structure of a leaf. Finally, they will study transport systems in both animals and plants by looking at the circulatory system and the transpiration stream.

Core concepts

- Cells to systems
- Photosynthesis

Prerequisite knowledge

- The hierarchical organisation of multicellular organisms: from cells to tissues to organs to systems to organisms
- The tissues and organs of the human digestive system, including adaptations to function and how the digestive system digests food (enzymes simply as biological catalysts)
- The role of diffusion in the movement of materials in and between cells
- The structure and functions of the gas exchange system in humans, including adaptations to function
- The mechanism of breathing to move air in and out of the lungs, using a pressure model to explain the movement of gases, including simple measurements of lung volume
- The reactants in, and products of, photosynthesis, and a word summary for photosynthesis
- The adaptations of leaves for photosynthesis
- The role of leaf stomata in gas exchange in plants

GCSE links

- Enzymes, and factors affecting the rate of enzymatic reactions
- Carbohydrates, proteins, nucleic acids, and lipids as key biological molecules
- The relationship between the structure and functions of the human circulatory system
- The need for transport systems in multicellular organisms, including plants
- The importance of cellular respiration; the processes of aerobic and anaerobic respiration
- Photosynthesis as the key process for food production and therefore biomass for life
- The process of photosynthesis

Key learning misconceptions

- Breathing and respiration are the same thing.
- Animals respire, plants do not (plants photosynthesise 'instead').
- Oxygen is not breathed out, only carbon dioxide (and vice versa – only oxygen is breathed in).
- Enzymes are living.
- Only organisms with lungs can respire.
- Plants obtain water through their leaves when it rains.
- Plant mass comes from the water and minerals from the soil.

Teaching preparation

Working scientifically

Model gut When revising the digestive system, ask students to imagine their body as having a hosepipe running from their mouth to their anus. This tube carries food and fluids into (and out of) the body. The analogy is limited, however, as unlike a hose the gut is permeable in parts so that nutrients and waste products can pass into the body from the tube, or from the body back into the tube. In this model, the villi would be tiny projections that stick into the tube to increase the surface area for absorption.

Key literacy skill

Enzyme names When introducing enzyme names, explain that the suffix *-ase* is used in biology to form the names of enzymes. The most common way to name enzymes is to add this suffix onto the end of the substrate – for example, an enzyme that breaks down carbohydrates is called carbohydrase. You could ask students to predict the names of other enzymes – for example, peroxides are broken down by peroxidase.

Working scientifically

Use of word equations to represent photosynthesis Link the use of word equations in biology to students' prior work in chemistry. Remind students that the reactants are always written on the left of the arrow, and the products on the right (the order on either side does not matter). In photosynthesis, 'light', which is required for the reaction but neither a reactant nor a product, is written above the arrow.

Key learning misconception

Plants do not respire Constantly reinforce the fact that plants also have to respire, to transfer the energy from the glucose produced in photosynthesis to the plant cells so that it can be used for growth. In B3 2.5 *Leaf structure and photosynthesis*, students look again at the role of stomata in gas exchange for photosynthesis. You may wish to explain to higher-ability students that, when a plant is photosynthesising, the net movement of gases is likely to be carbon dioxide into and oxygen out of the leaf. However, some molecules will move the other way as a result of respiration.

Key maths skill

Surface area to volume ratio Students will have been taught how to calculate the volume and surface area of common 3D shapes in maths, but you may need to recap these skills. To find volume, remind students to find the area of one face and then multiply this by the depth of the shape. (Note: some students may refer to this as height; in this case, volume = length × width × height.) Some students may find it easier to draw out the net of the shape when working out surface area. In this concept, limit the shapes to cubes and cuboids.

2.1 Reactivate: Cell systems

Themes: Bioenergetics, Organisation

Sub-themes: Circulatory system, Digestive system, Enzymes, Photosynthesis, Plant organ systems, Principles of organisation, Respiratory system

Biology NC links

- the tissues and organs of the human digestive system, including adaptations to function and how the digestive system digests food (enzymes simply as biological catalysts)
- the role of diffusion in the movement of materials in and between cells
- the structure and functions of the gas exchange system in humans, including adaptations to function
- the adaptations of leaves for photosynthesis
- the role of leaf stomata in gas exchange in plants

Learning objective

Recall and organise key knowledge about levels of organisation within an organism.

Year 7 learning objectives to revise

- Describe the adaptations of specialised plant cells
- Describe the process of diffusion
- Describe examples of diffusion
- Give the definitions of tissue, organ, and organ system
- Describe the hierarchy of organisation in a multicellular organism
- Describe how parts of the gas exchange system are adapted to their function

Year 8 learning objectives to revise

- Give the definition of an enzyme
- Describe the role of enzymes in carbohydrate, protein, and lipid digestion
- Describe the process of photosynthesis
- Give the word equation for photosynthesis
- Describe the main adaptations of a leaf
- Describe the role of stomata
- Describe how water is transported through a plant

Pinchpoint focus

The process of diffusion

Tier 2 vocabulary	Tier 3 vocabulary
cell, digestion, producers, specialised, transfer	*diffusion*, enzymes, *organ*, organ system, photosynthesis, small intestine, *tissue*, villi

Trigger interest

Show students an image of a person eating an apple. Ask them to list five key facts about digestion.

Show students an image of a leaf. Ask them to list five key features of a leaf.

Show students an image of a pair of scissors. Ask: *In what ways are the scissors similar to and different from an enzyme?*

Recall

Ask students to list the layers of organisation in the body, from smallest to biggest. They should then add an example of a structure at each level.

Ask students to write down the word equation for photosynthesis. Can students add a description of how the reactants of photosynthesis reach the cell?

Main activity

Pinchpoint: Diffusion (20–25 mins)

Diffusion is the process by which many substances, including oxygen and glucose, move into and out of cells. It is essential for respiration and photosynthesis.

Students answer the pinchpoint question and complete the follow-up activity with the same letter. They then complete the *What I need to remember* activity. Information about each incorrect answer's misconceptions can be found on the teacher sheet. You may want to divide the class into groups, depending on the letter answer chosen, to support each other when completing the follow-up activity.

Model diffusion using coloured ping pong balls. Distribute one set randomly in a glass trough of water, then add a second colour into a section of the trough (held in place with a hoop). Ask students where the highest concentration of each colour of particles is found. Disturb the water to represent random particle motion, then remove the hoop. Students should see the balls randomly spread out until there is an even distribution of balls throughout the liquid. Discuss what this represents in terms of concentration.

Students produce cartoon-style images, with captions, of three stages of the demonstration.

Students produce a poster describing the leaf's adaptations for photosynthesis, including a zoomed-in section of a stoma.

Review and reflect

Students are shown the pinchpoint question again and asked to reflect on whether they think they chose the correct letter and if they would change their answer now. If they think they got it wrong, ask them to think about why they chose that answer.

Reveal the correct answer. Review with students why it is correct and the other options are not. Guidance can be found on the teacher sheet. Students could be asked to pick an incorrect answer and write an explanation to a fictional student who chose it, explaining why it cannot be correct.

Students write a definition of diffusion and illustrate it using an example from digestion or photosynthesis. They must refer to the concentration of the molecule that is moving.

Homework

Students fill in the incomplete knowledge organiser.

2.2 Apply: Cells to organ systems

Theme: Organisation

Biology NC link

- the hierarchical organisation of multicellular organisms: from cells to tissues to organs to systems to organisms

Sub-themes: Circulatory system

Working scientifically NC link

- understand that scientific methods and theories developed as earlier explanations are modified to take account of new evidence and ideas, together with the importance of publishing results and peer review

Learning objective

- Use the circulatory system to illustrate the hierarchy of organisation in a multicellular organism.

Year 7 and 8 learning objectives to be applied

- Give the definitions of tissue, organ, and organ system
- Describe the hierarchy of organisation in a multicellular organism

Application

- Explain in detail how and why scientific methods or theories have changed over time (using the discovery of the circulatory system, from Galen to Harvey).

Tier 2 vocabulary	Tier 3 vocabulary
cell, hierarchy, organisation	circulatory system, multicellular, *organ*, organ system, organism, *tissue*

Answers to SB in-text questions

A Cells **B** Contract to cause movement **C** Organism

Lesson plan

Reactivate knowledge

1 What is an organ system?

2 Name some examples of organ systems.

3 What is a multicellular organism?

Lesson resources

Activity: *Organ systems* (information sheets × 2, student sheet, support sheet, teacher and technician sheet)

Answers: 1 A group of organs working together to perform a function **2** For example, digestive system, respiratory system, reproductive system **3** Organism made of more than one cell

Trigger interest

Show students a blank flow diagram to represent the levels of organisation in an organism. They then use the key terms to label the levels in the flow diagram and add examples. Ensure students can define each of the key terms – cell, tissue, organ, organ system, and organism.

Exposition of main content

Show students the images of organ systems on the Lesson presentation. Ask if they can recognise what they are by looking for key organs. If students are having difficulty, add the names of the organ systems to the board for them to choose from. Discuss the function of each system. With the exception of the circulatory and nervous systems, they have studied all the systems in some detail in Activate 1 and 2. You may wish to remind students that when studying movement, the skeleton and muscular system are referred to together as the musculoskeletal system.

Main activity

Activity: Organ systems (30 mins) Students read information about the circulatory system or the nervous system. They use this information to complete a flow diagram to illustrate the hierarchy of organisation in a multicellular organism. They then teach their partner about the organ system and complete a second flow diagram based on the system their partner learnt about. Finally, they practise answering a 6-mark exam-style question.

Discuss students' answers. Then explain how new technology and ideas lead to changes in our understanding of the way organisms work. Use the information in the Student Book on the work of Galen and Harvey to illustrate this. If time allows, ask students to carry out their own research into this area – for example, they could look at Hippocrates and the four humours.

Review and reflect

Name some structures in the circulatory system. Ask students to state which level of organisation they belong to and explain why.

Homework

Students answer the exam-style question:

Use the nervous system to illustrate the hierarchy of organisation in a multicellular organism. **(6 marks)**

Encourage students to use their nervous system flow diagram and circulatory system model answer to help them complete the task.

2.3 Apply: Digestive system and enzymes

Themes: Bioenergetics, Organisation

Sub-themes: Digestive system, Enzymes

Biology NC link

- the tissues and organs of the human digestive system, including adaptations to function and how the digestive system digests food (enzymes simply as biological catalysts)

Working scientifically NC links

- use appropriate techniques, apparatus, and materials during fieldwork and laboratory work, paying attention to health and safety
- make and record observations and measurements using a range of methods for different investigations; and evaluate the reliability of methods and suggest possible improvements
- interpret observations and data, including identifying patterns and using observations, measurements, and data to draw conclusions

Learning objective

- Apply the role of enzymes in carbohydrate, protein, and lipid digestion.

Year 7 and 8 learning objectives to be applied

- Give the definition of an enzyme
- Describe the role of enzymes in carbohydrate, protein, and lipid digestion

Application

- Enzymes used in commercial washing powder.

Tier 2 vocabulary	Tier 3 vocabulary
carbohydrate, commercially, evaluate, interpret, protein	amino acid, carbohydrase, *catalyst*, enzyme, fatty acid, glycerol, lipase, lipid, protease

Answers to SB in-text questions

A Small intestine **B** Baby foods and biological washing powders **C** Protease

Lesson plan

Reactivate knowledge

1 What is digestion?

2 What is an enzyme?

3 Name some digestive enzymes.

Lesson resources

Practical: *Investigating washing powders* (student sheet, support sheet, teacher and technician sheet)

Answers: 1 The process that breaks down large (insoluble) molecules into small (soluble) molecules **2** Substances that speed up reactions without being used up **3** Carbohydrases, lipases, proteases.

Trigger interest

Show students the image of clothes being washed. Ask: *What is the difference between biological and non-biological washing powders?* Explain that biological washing powders contain digestive enzymes that help to clean your clothes. Ensure students do not refer to enzymes as 'living things'.

Discuss with students what an enzyme is, and why it is referred to as a biological catalyst. Show students the blank table from the Student Book, then ask them to complete this, in pairs, matching the enzyme with the nutrient broken down and the product produced. This will allow you to assess what students can remember from Activate 2. If needed, write the terms on the board for students to choose from.

Exposition of main content

Using the information from the Student Book, introduce the commercial uses of enzymes in the production of baby food and washing powders.

Discuss the advantages and disadvantages of using biological washing powders.

Main activity

Practical: Investigating washing powders (25 mins)
Students investigate the effect of different washing powders and liquids on agar (contains carbohydrate) and gelatine (contains proteins) to determine which enzymes they contain. They are then provided with information on a commercial use of enzymes to interpret.

Review and reflect

Provide students with mini-whiteboards and pens. Read out a number of different foods that could have left marks on clothing. Students need to write down protease, lipase, and carbohydrase for the enzyme(s) that could be used to break each food down. For example, cheese contains protein and fats, so they should write down protease and lipase.

Homework

Students produce a graphic organiser summarising the uses of enzymes in biological washing powder, including the advantages and disadvantages of using biological washing powder.

2.4 Apply: Respiratory system and gas exchange

Theme: Organisation

Sub-theme: Respiratory system

Biology NC links

- the role of diffusion in the movement of materials in and between cells
- the structure and functions of the gas exchange system in humans, including adaptations to function
- the mechanism of breathing to move air in and out of the lungs, using a pressure model to explain the movement of gases, including simple measurements of lung volume
- the impact of exercise, asthma, and smoking on the human gas exchange system

Learning objective

- Explain how alveoli are adapted for gas exchange.

Year 7 and 8 learning objectives to be applied

- Describe the process of diffusion
- Describe examples of diffusion
- Describe how parts of the gas exchange system are adapted to their function

Application

- Features of exchange surfaces linked to factors that affect diffusion rate.

Tier 2 vocabulary	Tier 3 vocabulary
absorb, distance, volume	alveoli, concentration gradient, *diffusion*, exchange surface, surface area, villi

Answers to SB in-text questions

A For aerobic respiration/to transfer energy to the cell **B** The villi in the walls of the small intestine **C** Thin walls, large surface area, good blood supply

Lesson plan

Reactivate knowledge

1 What is diffusion?
2 What factors affect the rate of diffusion?
3 Where in the body does oxygen diffuse into the blood?

Lesson resources

Activity: *Exchange surfaces* (information sheets × 3, student sheet, support sheet, teacher and technician sheet)

Answers: 1 Movement of particles from an area of high concentration to an area of low concentration **2** Concentration gradient, temperature, surface area **3** Through the alveoli in the lungs

Trigger interest

Show students the image of the flatworm in Figure 1 from the Student Book. Ask: *Do you know what it is? How would you describe its shape?* Explain that it is a flatworm. By being so flat and thin, the animal can take in all the essential substances it needs (such as oxygen and nutrients) from its environment through its skin.

Exposition of main content

Explain that the surface area of an organism defines how quickly it can absorb substances from its environment, and an organism's volume determines how much of a substance it needs. Small organisms such as flatworms exchange essential and waste substances through their body surfaces by diffusion, as they have a large surface area to volume ratio.

Check students' understanding of the terms 'surface area' and 'volume' by showing them a range of shapes such as a cube and a sphere. Then use Figure 2 from the Student Book to show students how to calculate the surface area to volume ratio of a cube, using the 1 cm cube example. Ask students to calculate the surface area to volume ratio of the 3 cm cube (and a 5 cm cube, if further practice is needed). Ask them to suggest what would happen if they calculated the surface area to volume ratio of a bigger cube. Explain that the larger the organism, the lower the surface area to volume ratio becomes. This makes it harder for substances to diffuse directly into the organism. Most multicellular organisms cannot use simple diffusion to survive, as diffusion cannot occur quickly enough to meet every cell's demands.

Explain that multicellular organisms get over these problems by having exchange surfaces such as the alveoli in the lungs for gas exchange and the villi in the small intestine for absorbing nutrients. Discuss the key features of an exchange surface.

Main activity

Activity: Exchange surfaces (20 mins) Students visit different stations in the classroom where they find out about three exchange surfaces: the alveoli in the lungs, the villi in the intestine, and the underside of a leaf. They learn how each structure is adapted to perform its function and record this in a table. Make sure that students do not confuse the walls in aveoli and villi with cell walls.

Review and reflect

Students produce a Venn diagram to compare the structure and function of the alveoli and the villi.

Homework

Students choose either an alveolus or a villus and make a model of its structure. They add labels to name the key parts and an explanation of how these features maximise the rate of exchange of named substances.

2.5 Apply: Leaf structure and photosynthesis

Themes: Bioenergetics, Organisation

Sub-themes: Photosynthesis, Plant organ systems

Biology NC links

- the reactants in, and products of, photosynthesis, and a word summary for photosynthesis
- the adaptations of leaves for photosynthesis
- the role of leaf stomata in gas exchange in plants

Working scientifically NC links

- use appropriate techniques, apparatus, and materials during fieldwork and laboratory work, paying attention to health and safety
- make and record observations and measurements using a range of methods for different investigations; and evaluate the reliability of methods and suggest possible improvements

Learning objective

- Explain how the structure of a leaf is adapted for photosynthesis.

Year 7 and 8 learning objectives to be applied

- Describe the process of photosynthesis
- Give the word equation for photosynthesis
- Describe the main adaptations of a leaf
- Describe the role of stomata
- Describe how water is transported through a plant

Application

- Structure of a leaf.

Tier 2 vocabulary	Tier 3 vocabulary
cross section	chlorophyll, *diffusion*, epidermis, palisade mesophyll, phloem vessel, photosynthesis, spongy mesophyll, stomata, xylem vessel

Answers to SB in-text questions

A Osmosis (or diffusion) **B** To absorb as much light as possible for photosynthesis **C** Waxy layer, upper epidermis, palisade mesophyll, spongy mesophyll, lower epidermis (contains stomata)

Lesson plan

Reactivate knowledge

1 What type of organisms carry out photosynthesis?
2 How does a plant take in water?
3 What do stomata do?

Lesson resources

Practical: *Observing a leaf* (student sheet, support sheet, teacher and technician sheet)

Answers: 1 Plants and algae **2** Osmosis through the root hair cells (from the soil) **3** Allow gases to diffuse into and out of the leaf

Trigger interest

Prompt students to recall how water travels to the leaves from the root hair cells. This is a good opportunity to check that students remember the differences between the structure and function of the phloem and the xylem vessels, as well as to link back to tissues. Explain that both of these vessels are found within a plant vein.

Exposition of main content

Recap the general adaptations of a leaf for photosynthesis: they are green as they contain chlorophyll to absorb light, have a large surface area to maximise light absorption, are thin to provide a short diffusion distance for carbon dioxide to move into the leaves, and have veins to transport the water.

Explain to students that although leaves are thin, they are made up of a number of layers. Introduce the structure of the leaf by showing them the leaf cross section in the Student Book. Explain that looking at a cross section of a leaf is like slicing a cake to see the layers. Talk through each layer of the leaf, starting with the waxy cuticle and working down through the layers. If possible, have some waxy leaves, such as privet, for students to feel. Students should be familiar with the role of guard cells in opening and closing stomata from Activate 2.

Main activity

Practical: Observing a leaf (30 mins) Students will look through a microscope at a cross section through a leaf, identify the tissues present, and draw the cross section. They will then answer the questions that follow.

Review and reflect

State an adaptation or function of a part of the leaf. On mini-whiteboards, students write the structure that is being referred to, and hold it up.

Homework

Students produce a visual summary to explain how a leaf is adapted to maximise photosynthesis.

2.6 Reach: Circulatory system

Theme: Organisation

Sub-theme: Circulatory system

Biology NC links

- the hierarchical organisation of multicellular organisms: from cells to tissues to organs to systems to organisms (KS3)
- the relationship between the structure and functions of the human circulatory system (KS4)

Learning objective

- Describe the structure and function of the circulatory system.

Year 7 and 8 learning objective to be built on

- Give definitions of tissue, organ, and organ system

Key Stage 4 link

- The relationship between the structure and functions of the human circulatory system.

Tier 2 vocabulary	Tier 3 vocabulary
	artery, atrium, capillary, cardiac muscle, double circulatory system, lumen, valve, vein, ventricle

Answers to SB in-text questions

A Vein **B** Lungs to heart to rest of the body to heart to lungs

Lesson plan

Reactivate knowledge

1 What is an organ system?

2 What is the function of the circulatory system?

3 Name some structures in the circulatory system.

Answers: 1 Group of organs working together to perform a function **2** Transport substances around the body **3** Heart, blood, blood vessels

Lesson resources

Activity: *Circulatory system* (student sheet, support sheet, teacher and technician sheet)

Trigger interest

Show students the image of the blocked artery in Figure 1 of the Student Book. Ask: *What do you think this is? What problems would this cause?* Explain that it is a blood vessel leading to the heart, and it is clogged with fat. Blood and therefore oxygen cannot reach the heart. This could stop the heart beating. Ask: *Can you name any of the different types of blood vessel?* Be aware students may have experience of friends/relatives who have suffered heart attacks.

Exposition of main content

Using Figures 2, 3, and 4 from the Student Book, introduce the three different types of blood vessel, talking through their function and structure. To help students remember the difference between the functions of arteries and veins, share the tip that artery begins with an 'a', which stands for carrying blood <u>a</u>way from the heart; vein contains the word 'in' which means that veins carry blood <u>in</u>to the heart.

Introduce simply the structure of the circulatory system using Figure 5 in the Student Book. Explain that humans (and most animals) have a double circulatory system – blood flows through the heart twice during each circuit of the body. Explain that blood is not really blue but this is the convention for drawing blood that contains low levels of oxygen. (Students may ask why veins appear to be blue. Explain that the whole light spectrum is incident on your skin. Most of the red portion of white light scatters off the skin before the light reaches the veins. The remaining light that reflects off the veins back into our eyes is mainly the blue portion of white light, so veins appear blue.)

Using Figure 6 in the Student Book, point out the main structures in the heart: it is made of cardiac muscle and has four chambers – two atria at the top and two ventricles at the bottom. Reinforce the fact that blood enters the heart from the veins and leaves through the arteries, and the presence of valves prevents the backflow of blood. Explain that the left side of the heart is drawn on the right on a diagram, like looking in a mirror. The artery leading to the lungs (pulmonary artery) contains blood with low oxygen and the vein returning to the heart (pulmonary vein) is oxygen-rich – usually, arteries contain oxygen-rich blood and veins contain oxygen-poor blood, but the pulmonary artery and pulmonary vein are exceptions.

Main activity

Activity: Circulatory system (30–35 mins) This activity is split into two tasks. In the first (shorter) task, students calculate their mean heart rate. In the second (longer) task, they make models of blood vessels using a range of materials.

Review and reflect

Provide students with mini-whiteboards and pens. Read out a series of statements about the structure or function of a blood vessel. Students write down the corresponding type of blood vessel (artery, vein, or capillary) and hold up their answers.

Homework

Students write an exam-style question and associated mark scheme, worth 4–6 marks, on the structure of the circulatory system or the heart.

2.7 Reach: Transpiration

Theme: Organisation

Biology NC links

- the hierarchical organisation of multicellular organisms: from cells to tissues to organs to systems to organisms (KS3)
- the role of leaf stomata in gas exchange in plants (KS3)
- the need for transport systems in multicellular organisms, including plants (KS4)

Sub-theme: Plant organ systems

Working scientifically NC links

- use appropriate techniques, apparatus, and materials during fieldwork and laboratory work, paying attention to health and safety
- make and record observations and measurements using a range of methods for different investigations; and evaluate the reliability of methods and suggest possible improvements

Learning objective

- Describe the transpiration stream.

Year 7 and 8 learning objectives to be built on

- Describe the adaptations of specialised plant cells
- Describe the role of stomata
- Describe how water is transported through a plant

Key Stage 4 link

- Movement of water through the plant in the xylem and loss of water through the stomata.

Tier 2 vocabulary	Tier 3 vocabulary
constant, evaporate	*diffuse*, guard cell, stomata, transpiration, transpiration stream, xylem vessel

Answers to SB in-text questions

A Root hair cell **B** (Soil), roots/root hair cells, xylem vessel in stem, xylem vessel in leaf, leaf cell, air space in leaf, (atmosphere) **C** Closed to prevent water loss and there is no light/poor conditions for photosynthesis

Lesson plan

Reactivate knowledge

1 What does the xylem do?
2 What does the phloem do?
3 What do xylem vessels look like?

Lesson resources

Practical: *Observing transpiration* (student sheet, support sheet, teacher and technician sheet)

Answers: 1 Transports water and mineral ions **2** Transports dissolved food/sugars **3** Long hollow tubes with lignin spirals in cell wall

Trigger interest

Show students Figure 1 from the Student Book (or a real-life example, if possible) of a flower whose stem has been cut in half and placed in two beakers of different-coloured water. Ask students to describe what they can see. Ask: *How do you think this has happened?*

Explain that the dye is taken into the stem and then transported around the plant as a result of transpiration.

Exposition of main content

Show students the transpiration stream in Figure 2 in the Student Book. Point out the movement of water through the roots, up the stem, into the leaf, and out of the plant.

Then, using the more detailed diagram in Figure 3, explain that transpiration is the process by which water is lost from the leaves (water evaporates into the air space in the spongy mesophyll, and the water vapour then diffuses out of the stomata). This loss of water from the top of the plant reduces the pressure at the top of the xylem vessels, causing water to move from high pressure at the bottom of the tube to low pressure at the top.

Use two long balloons in conjunction with Figure 4 in the Student Book to represent how the guard cells control the opening and closing of the stomata. When you inflate the balloons, like the guard cells filling up with water, they become rigid and open the stomata. When you deflate the balloons, like the guard cells losing water, they become floppy and the hole closes over.

Main practical

Practical: Observing transpiration (30 mins) In this practical, students observe the xylem vessels in celery that has been soaked overnight in dye. They then answer questions on the transpiration stream and the role of stomata in controlling water loss.

As an alternative extension to the suggestion in the Lesson presentation, students could be asked how they would improve the accuracy of their answer to how far the dye travelled up the celery. (They were only checking every 5 cm, so if they checked in smaller increments they would get more accurate data.)

Review and reflect

Provide students with the terms below to arrange into a flow diagram to represent the movement of water through the transpiration stream:

stomata palisade leaf cell spongy mesophyll
xylem vessel root hair cell

Can they add the correct term 'diffusion' or 'evaporation' over each step of the flow diagram?

Homework

Students sketch a plant and use this to produce a visual summary of the transpiration stream. For extension, ask them to zoom in on the leaf to explain how the opening and closing of the stomata controls the loss of water vapour.

Introduction to checkpoint intervention

This checkpoint intervention provides suggestions for a lesson to follow up the B3 2 *Cells systems* checkpoint assessment. Depending on the outcomes of the assessment, these suggestions could move a class towards secure, to target specific learning outcomes, or to consolidate knowledge for students achieving secure outcomes. Use students' outcomes from the checkpoint assessment to address general misconceptions from the content covered. Be prepared to re-cover content as required; three key concepts from the chapter are covered in more detail below.

Checkpoint secure learning outcomes

- Use the circulatory system to illustrate the hierarchy of organisation in a multicellular organism.
- Interpret information on the commercial use of digestive enzymes.
- Explain how alveoli are adapted for gas exchange.
- Explain how the structure of a leaf is adapted for photosynthesis.
- Describe the structure and function of the circulatory system.
- Describe the transpiration stream.

Student reflection

Encourage students to reflect on whether there were any checkpoint questions they found difficult or straightforward, as well as their preparation for the checkpoint assessment, including how they revised, the time spent on revision, and what areas they could improve on. For more strategies, see the Metacognition chapter or the Metacognition in Key Stage 3 Science guide on Kerboodle.

Intervention activity: Interpret information on the commercial use of digestive enzymes

Students often have difficulty applying their knowledge to unfamiliar contexts, especially where new information is also presented. For example, the addition of enzymes to baby foods to help break down proteins relies on prior knowledge, but introduces a context that may be unfamiliar to students. Asking students to write down what they know about a topic before they start trying to answer the question will help them to understand what they are being asked to do.

Digestive enzymes

Begin this activity by asking students to work in pairs to produce a summary table that contains the names of the digestive enzymes, what they break down, and what they produce. They should then refer to this table before attempting questions that require them to apply their knowledge of enzymes. Examples of questions could include:

1. Blood contains proteins. Which enzyme should you use to remove a blood stain? Explain why.
2. Bread contains starch. Which enzyme could be added to bread to make it taste sweeter? Explain your answer.

For support, students could refer to the information in B3 2.3 *Digestive system and enzymes*.

Intervention activity: Explain how alveoli are adapted for gas exchange

It is important to reiterate to students the difference between the command words 'describe' and 'explain'. In this example, they need to link the alveoli adaptations to how the efficiency of gas exchange is increased. Students often find it useful to underline key parts of the question, such as 'explain', 'alveoli', and 'gas exchange', before attempting to answer the question. Students requiring support may need prompts, key words, or sentence starters to help structure an answer appropriately.

Explaining the adaptations of alveoli

Provide students with the following exam-style question and student answer. Ask them to mark it.

Explain ***two*** *ways the alveoli are adapted to their function.* **(4 marks)**

The role of the gas exchange system is to take oxygen into the body and remove carbon dioxide.

Gas exchange takes place in the lungs, which contain many alveoli.

The alveoli have very thin walls.

This provides a short path for oxygen to diffuse through.

They also have a good blood supply.

Students should have scored the question three marks. The answer has named two features but only explained the first one. The first and second lines are not markworthy.

You could ask students to write a model answer for two different adaptations in the gas exchange system. For support, students could refer to the information in B3 2.4 *Respiratory system and gas exchange*.

Intervention activity: Explain how the structure of a leaf is adapted for photosynthesis

Students often find it easy to describe some ways that a leaf is adapted to photosynthesis but do not always back this up with an explanation. For example, a common answer is that 'Leaves are green and flat'. Stomata are also frequently mentioned but not the epidermis or the air spaces in the spongy mesophyll.

Leaf structure

Working as a whole class, ask students to identify all the main structures in the leaf and discuss the role they play in photosynthesis. Students produce a poster or model of the structure of a leaf with all the key features labelled. They then add descriptions explaining how the various structures enable the leaf to maximise photosynthesis. For support, students could refer to the information in B3 2.5 *Leaf structure and photosynthesis*.

B3 Chapter 3: Fertilisation and implantation

Introduction to chapter

In this chapter, students will compare sexual reproduction in plants and animals. They will look at pregnancy prevention using contraception before interpreting menstrual cycle diagrams. They will investigate factors that affect seed dispersal and the steps that should be taken to collect valid data. Finally, they will discover how different alleles affect which characteristic is most likely to be displayed in an organism's offspring and how some organisms reproduce asexually.

Core concepts

- Fertilisation and implantation

Prerequisite knowledge

- Reproduction in humans (as an example of a mammal), including the structure and function of the male and female reproductive systems, menstrual cycle (without details of hormones), gametes, fertilisation, gestation, and birth, to include the effect of maternal lifestyle on the foetus through the placenta
- Reproduction in plants, including flower structure, wind and insect pollination, fertilisation, and seed and fruit formation and dispersal, including quantitative investigation of some dispersal mechanisms
- Heredity as the process by which genetic information is transmitted from one generation to the next

GCSE links

- Single gene inheritance and single gene crosses with dominant and recessive phenotypes
- Genetic variation in populations of a species
- The uses of modern biotechnology including gene technology; some of the practical and ethical considerations of modern biotechnology

Key learning misconceptions

- Pollination and fertilisation are the same process.
- Plants do not reproduce sexually.
- Pollen grains move down the pollen tube.
- Bees are the only insects that pollinate flowers and all flowers are pollinated by bees.
- Adolescence occurs in the teenage years and ends at either 18 or 21 years.
- Periods start only in your teenage years.
- All methods of contraception are 100% effective.
- The uterus is lost during menstruation.
- You can't get pregnant the first time you have sex or when standing up.
- Characteristics are all caused by genetics.

Teaching preparation

Key vocabulary

Scientific names The use of many key scientific terms dominates this topic, such as the names of the parts of a flower and the male and female reproductive systems. Encourage students to make their own scientific dictionary of terms, perhaps at the back of their exercise book. They can add words throughout the topic and then refer to these when answering any questions. Encourage regular practice of writing the words and saying them out loud.

Key learning misconception

Reproduction Students often have many queries/misconceptions regarding adolescence, pregnancy, and contraception. Allow time in these lessons for students to ask appropriate questions in a safe environment. For example, in the form of a 'questions box' where students are able to pose questions anonymously, which you can then read through and choose to answer in the next lesson. Be mindful to constantly reinforce that changes occur at different times for different individuals.

Key vocabulary

Sex and gender In this topic, students revisit sexual reproduction, the menstrual cycle, and techniques for preventing pregnancy. Throughout the Student Book, the person's biological sex is referenced. Gender is the term used more broadly to denote a range of identities which may or may not correspond to the biological sex classifications of male and female, which are based on reproductive functions (whether they produce egg cells or sperm cells).

Someone who identifies as cisgender is a person whose gender identity is the same as the sex that was registered for them at birth. Trans is a term for people whose gender is not the same as the sex that was registered for them at birth, and who may or may not wish to identify with a particular gender. Trans people may use other terms, including transgender, genderqueer, gender-fluid, non-binary, and agender.

Binary is a word used to refer to something relating to two things. Binary gender identities are male or female. Non-binary is a term for people who don't agree with the identity of being exclusively either female or male. Non-binary identities can include people who adopt some aspects of male/female identities or people who don't identify with a binary gender at all.

3.1 Reactivate: Fertilisation and implantation

Themes: Genetics and inheritance, Reproduction

Sub-themes: Asexual reproduction, Human reproductive systems, Inheritance and DNA structure, Menstrual cycle and hormonal control of reproduction, Plant reproduction

Biology NC links

- reproduction in humans (as an example of a mammal), including the structure and function of the male and female reproductive systems, menstrual cycle (without details of hormones), gametes, fertilisation, gestation and birth, to include the effect of maternal lifestyle on the foetus through the placenta
- reproduction in plants, including flower structure, wind and insect pollination, fertilisation, seed and fruit formation and dispersal, including quantitative investigation of some dispersal mechanisms

Learning objective

Recall and organise key knowledge about reproduction and inheritance.

Year 7 learning objectives to revise

- Describe the biological function of the main structures in the female and male reproductive systems
- Describe the process of fertilisation
- Describe the main stages in the menstrual cycle
- Describe some different methods of contraception
- Describe the function of the main structures in a flower
- Describe the process of fertilisation in plants
- Explain how seeds are adapted for their method of dispersal
- Describe a method to investigate seed dispersal

Year 8 learning objectives to revise

- Describe how characteristics are inherited
- Give the definitions of DNA, chromosome, and gene

Pinchpoint focus

Process of fertilisation in both plants and animals

Tier 2 vocabulary	Tier 3 vocabulary
characteristic, condom, contraception, cycle, inheritance, pollen, variation	anther, carpel, chromosome, contraceptive pill, *DNA* fertilisation, filament, gene, menstrual cycle, ovary, oviduct, ovulation, ovule, penis, pollination, stamen, stigma, style, testis, uterus, vagina

Trigger interest

Show an image of an embryo/ early foetus and a plant seed. Ask students what these structures have in common and how the processes by which they have formed are similar.

Show students an image of DNA. Ask them to recognise what the picture is of and state three facts about it.

Show students an image of a range of different methods of contraception. Ask them to describe their function and list as many forms of contraception as possible.

Recall

Ask students to draw a diagram of a cross-section through a flower, labelling as many structures as possible. Students should then add descriptions for each part.

List the main stages in the menstrual cycle. Where possible, add a time reference and any additional information.

Main activity

Pinchpoint: Fertilisation (20–25 mins)

Fertilisation is an essential process in both plants and animals in the formation of offspring. Multiple steps have to take place for this to occur, which students need to be aware of. This knowledge also forms the basis of understanding variation.

Students answer the pinchpoint question and complete the follow-up activity with the same letter. They then complete the *What I need to remember* activity. Information about each incorrect answer's misconceptions can be found on the teacher sheet. You may want to divide the class into groups, depending on the letter answer chosen, to support each other when completing the follow-up activity.

Students produce a flow diagram, showing the processes of pollination and fertilisation in a plant to form a seed. They should include a sentence describing each stage on their diagram.

In pairs, students produce a 'key words' quiz on plant and animal reproductive structures and processes. One produces the animal quiz, the other the plant quiz. They then use their quiz to test their partner.

Review and reflect

Students are shown the pinchpoint question again and asked to reflect on whether they think they chose the correct letter and if they would change their answer now. If they think they got it wrong, ask them to think about why they chose that answer.

Reveal the correct answer and review with students why it is correct and the other options were not. Guidance can be found on the teacher sheet. Students could be asked to pick an incorrect answer and write an explanation to a fictional student who chose it, explaining why it cannot be correct.

Students produce a table comparing fertilisation in plants and animals. Headings could include the names of the gametes, where they are produced, and where fertilisation takes place.

Describe some key reproductive structures or processes. Students write the key word on their mini-whiteboards and hold it up.

Homework

Students fill in the incomplete knowledge organiser.

3.2 Apply: Sexual reproduction

Theme: Reproduction

Sub-themes: Human reproductive systems, Plant reproduction

Biology NC links

- reproduction in humans (as an example of a mammal), including the structure and function of the male and female reproductive systems, menstrual cycle (without details of hormones), gametes, fertilisation, gestation and birth, to include the effect of maternal lifestyle on the foetus through the placenta
- reproduction in plants, including flower structure, wind and insect pollination, fertilisation, seed and fruit formation and dispersal, including quantitative investigation of some dispersal mechanisms

Learning objective

- Compare the process of sexual reproduction in plants and animals.

Year 7 and 8 learning objectives to be applied

- Describe the biological function of the main structures in the female and male reproductive systems
- Describe the process of fertilisation
- Describe the function of the main structures in a flower
- Describe the process of fertilisation in plants
- Describe how characteristics are inherited

Application

- Features of sexual reproduction and comparison of the process of fertilisation in plants and animals.

Tier 2 vocabulary	Tier 3 vocabulary
fertilisation, identical, pollen, variation	egg cell, gamete, ovule, pollination, sexual reproduction, sperm cell

Answers to SB in-text questions

A Each sperm cell and egg cell contains one copy of each chromosome. During fertilisation, the genetic material joins together. The fertilised egg divides several times to form an embryo. Each cell in the embryo contains two copies of each chromosome **B** A pollen grain is transferred from the anther of one flower to the stigma of another. The pollen grain nucleus travels down the pollen tube to an ovule, where it joins with the ovule nucleus to create a seed **C** Similarity: both require two parents. Difference: the male gamete in a plant is a pollen grain; in an animal, it is a sperm cell

Lesson plan

Reactivate knowledge

Lesson resources

Activity: *Sexual reproduction and fertilisation* (student sheet, teacher and technician sheet)

1. What are gametes?
2. Name the two animal gametes.
3. Name the two plant gametes.

Answers: 1 Sex cells **2** Sperm and egg cell **3** Pollen and ovule (egg) cell

Trigger interest

Ask students to write down five facts about sexual reproduction, then share their ideas. Did any students refer to plants? Remind students that sexual reproduction refers to the mixing of gametes, whereas sexual intercourse is the process by which sperm are transferred from the penis of the male into the vagina of the female.

Exposition of main content

Explain that sexual reproduction has three key features: two parents are needed, genetic material is mixed during fertilisation, and variation is produced in the offspring. Use Figure 1 in the Student Book to recap the process of fertilisation in animals, including where the gametes are made in the body and where fertilisation occurs.

Then discuss the process of sexual reproduction in plants, using Figure 3 in the Student Book to recap the process of fertilisation. Highlight similarities and differences in the structures involved and the process of fertilisation between plants and animals. Ask: *How is pollination similar to sexual intercourse?*

Main activity

Activity: Sexual reproduction and fertilisation (35–40 mins) This activity is split into two tasks. In Task 1, students are split into groups and make a poster on sexual reproduction in either animals or plants, then share information with another group. In Task 2, students use the *comprehension, connection, strategies, and reflection* metacognition activity to answer the following exam-style question:

Compare the processes of sexual reproduction in plants and animals. **(6 marks)**

Review and reflect

Ask students to list the three key features of sexual reproduction.

Homework

Students produce a Venn diagram to compare the similarities and differences between fertilisation in plants and animals.

3.3 Apply: Preventing pregnancy

Theme: Reproduction

Sub-themes: Human reproductive systems, Menstrual cycle and hormonal control of reproduction

Biology NC link

- reproduction in humans (as an example of a mammal), including the structure and function of the male and female reproductive systems, menstrual cycle (without details of hormones), gametes, fertilisation, gestation and birth, to include the effect of maternal lifestyle on the foetus through the placenta

Learning objective
• Evaluate the use of some different methods of contraception.

Year 7 and 8 learning objectives to be applied	Application
• Describe the biological function of the main structures in the female and male reproductive systems • Describe the main stages in the menstrual cycle • Describe some different methods of contraception	• Explaining how different forms of contraception work

Tier 2 vocabulary	Tier 3 vocabulary
contraception, effectiveness, evaluate, fertilisation, hormone, implantation, pregnancy, prevention, protection	egg cell, embryo, ovary, oviduct, seed, sperm cell, spread, uterus

Answers to SB in-text questions

A Order: fertilisation, implantation, pregnancy, sexual reproduction **B** To kill sperm (to prevent an unwanted pregnancy)

Lesson plan

Reactivate knowledge

1 What happens during fertilisation in animals?

2 Name the organ in a female's body where/into which sperm is released during sexual intercourse.

3 Give some examples of contraception.

Answers: 1 Nucleus of the sperm and egg join together **2** Vagina **3** For example, condom, contraceptive pill

Lesson resources

Activity: *Contraception* (student sheet, support sheet, teacher and technician sheet)

Trigger interest

Ask students to define what contraception is, and then name as many forms of contraception as they can. Use Figure 1 in the Student Book to prompt thinking, if necessary.

Exposition of main content

Assess prior knowledge by using students' ideas to produce a flow diagram on the board; it should show the steps that must occur for a female to become pregnant. Use Figure 2 in the Student Book as a guide, if necessary.

Explain that there are three methods of contraception – barrier, chemical, and hormonal. Discuss some advantages and disadvantages of each method (students will discover more when they carry out their own research), showing Figure 1 from the Student Book.

Finally, discuss the effectiveness of the different methods. Explain that no form of contraception is 100% effective. Show Figure 3 from the Student Book to illustrate this. Students can then complete the task in the Reading graphs margin box.

Main activity

Activity: Contraception (35–40 mins) Students carry out their own research into different forms of contraception. They then design a leaflet about contraception. The leaflet will be given out at doctors' surgeries, and should provide general advice about contraception and the options available.

Review and reflect

Hand out mini-whiteboards and marker pens to students. State a type of contraceptive. Students should write down whether the hormonal, chemical, or barrier method(s) of contraception would be appropriate to the named type of contraception. For example, most condoms would be classified as both a barrier and a chemical method.

Homework

Students produce a graphic organiser to summarise the advantages and disadvantages of each type of contraceptive, and describe how each one works.

3.4 Apply: The menstrual cycle

Theme: Reproduction

Sub-theme: Menstrual cycle and hormonal control of reproduction

Biology NC link

- reproduction in humans (as an example of a mammal), including the structure and function of the male and female reproductive systems, menstrual cycle (without details of hormones), gametes, fertilisation, gestation and birth, to include the effect of maternal lifestyle on the foetus through the placenta

Working scientifically NC link

- interpret observations and data, including identifying patterns and using observations, measurements, and data to draw conclusions

Learning objective

- Interpret diagrams showing the main stages in the menstrual cycle.

Year 7 and 8 learning objectives to be applied

- Label the main structures in the female and male reproductive systems
- Describe the biological function of the main structures in the female and male reproductive systems
- Describe what happens during a period
- Describe the main stages in the menstrual cycle

Application

- Analysing a graph to show how LH and oestrogen hormones control the menstrual cycle.

Tier 2 vocabulary	Tier 3 vocabulary
hormone, puberty, typical	lutenising hormone, menstrual cycle, oestrogen, ovary, ovulation, *period*, uterus

Answers to SB in-text questions

A Four days **B** Day 12–16 **C** For example, release of egg cells, pubic hair growth, emotional changes **D** Increase

Lesson plan

Reactivate knowledge

1 What happens during a female's period?
2 What is ovulation?
3 Approximately how long does the menstrual cycle last?

Answers: 1 Uterus lining is shed **2** Egg is released from an ovary **3** 28 days/one month

Lesson resources

Activity: *Hormonal control of the menstrual cycle* (student sheet, support sheet, teacher and technician sheet)

Trigger interest

Show students the female reproductive system in Figure 1 from the Student Book. Ask them to identify the structures and describe their function. Point out the lining of the uterus and discuss what happens to it during a female's period. It is important that students always refer to the **lining** of the uterus being lost, rather than the uterus itself.

Exposition of main content

Talk through the main stages in the menstrual cycle, using Figure 2 in the Student Book. Ask students to read off dates from the diagram which correspond to the particular events. You may wish to discuss how the diagram would be different if the female became pregnant.

Discuss what is meant by a hormone. Start by explaining the function of oestrogen and its role in puberty, with which students should be familiar. Then introduce its role in the menstrual cycle, using Figure 3 in the Student Book to illustrate how the level of the hormone causes an event. For example, as oestrogen levels rise, the uterus lining builds up. Then do the same for LH. (At this stage, students do not need to be aware of how FSH and progesterone interact with LH and oestrogen to control the menstrual cycle.)

Main activity

Activity: Hormonal control of the menstrual cycle (25 mins) This activity is split into two tasks. In Task 1, students annotate diagrams to show what is happening to the uterus and the egg at different stages in the menstrual cycle. In Task 2, students interpret a graph showing how hormones control the menstrual cycle.

Review and reflect

Students sketch a graph to show the levels of oestrogen and LH throughout the menstrual cycle. They then add a sentence at different points on the graph to explain what each hormone is doing; for example: *at the highest level of LH, ovulation occurs.*

Homework

Students produce a timeline to show the key stages in the menstrual cycle. They add labels to describe what is happening to levels of LH and oestrogen at different points in the cycle.

3.5 Apply: Investigating seed dispersal mechanisms

Theme: Reproduction

Sub-theme: Plant reproduction

Biology NC link

- reproduction in plants, including flower structure, wind and insect pollination, fertilisation, seed and fruit formation and dispersal, including quantitative investigation of some dispersal mechanisms

Working scientifically NC links

- select, plan, and carry out the most appropriate types of scientific enquiries to test predictions, including identifying independent, dependent, and control variables
- use appropriate techniques, apparatus, and materials during fieldwork and laboratory work, paying attention to health and safety
- make and record observations and measurements using a range of methods for different investigations; and evaluate the reliability of methods and suggest possible improvements

Learning objective

- Explain the steps taken to produce valid data in a seed dispersal investigation.

Year 7 and 8 learning objectives to be applied

- Describe the advantages of seed dispersal
- Explain how seeds are adapted for their method of dispersal
- Describe a method to investigate seed dispersal

Application

- Factors that affect seed dispersal.

Tier 2 vocabulary	Tier 3 vocabulary
evaluate, fertlisation, reproducible, spread, valid	adaptation, pregnancy, seed dispersal

Answers to SB in-text questions

A Seeds **B** The variability in results **C** Data are reproducible if the investigation is repeated by another person, or by using different equipment, and the same results/data are obtained

Lesson plan

Reactivate knowledge

1 What is meant by seed dispersal?

2 Name some ways seeds are dispersed.

3 Give some examples of seed dispersal adaptations.

Lesson resources

Practical: *Seed dispersal investigation* (student sheets × 2, support sheet, teacher and technician sheet)

Answers: 1 Mechanism by which seeds are spread away from the parent plant **2** For example, wind, water, animals, explosion **3** For example, hooks, wings, light mass

Trigger interest

Show students the image of seeds in Figure 1 from the Student Book. Ask: *What does the image show? Can you list five facts about the structures?* Share students' ideas to assess their prior knowledge about seeds. You may wish to show the structure of the seed in Figure 2 when discussing the function of different parts of the seed. This is a good time to reiterate the fact that seeds do not need light to germinate. To start with, they gain all the energy they need to grow from their food store.

Discuss why seeds are dispersed and the different mechanisms of seed dispersal. Suitable examples are summarised in Table 1 in the Student Book.

Exposition of main content

Show students the photo of the sycamore seed in Figure 2 in the Student Book. Ask students to explain how it is adapted to being dispersed by the wind (low mass and has wings). Then use the example to promote discussion about what factors students could investigate that may affect how far the seed travels. Collate a list of factors on the board.

Explain that students are going to design their own investigation into seed dispersal mechanisms, and they need to ensure that they collect valid data. Explain that data from a scientific investigation are valid if the method ensures that values of the dependent variable are affected only by the independent variable. Therefore, students must ensure that all other variables are controlled effectively.

Then show the graph in Figure 3 from the Student Book. Explain that, if data show a number of outliers, or repeat data contain a large spread, or data are not reproducible, then the data may not be valid.

Explain the importance of evaluating an experimental method when determining how confident students can be about their conclusions. If they suspect that the method has produced invalid data, then conclusions are only tentative.

Main activity

Practical: Seed dispersal investigation (35–40 mins)
Students design their own investigation into seed dispersal mechanisms. They focus on collecting valid data. After collecting their data and analysing their results, they evaluate their method to decide whether the data collected are valid.

A guided investigation into seed dispersal was included in *Activate 1*. Therefore, you may wish to dissuade students from investigating wing length. Suitable variables students could investigate are: wing speed (using different fan speed settings), the height from which the seed is dropped, and mass of seed (mass can be altered by adding different numbers of paperclips to the seeds).

Review and reflect

Students reflect on their practical using the questions on the Lesson presentation. Then share students' ideas. When thinking about outliers, remind students to look at their graphs: *Are the data close to the line of best fit?* The spread of data can be obtained by looking at repeat results: *Are these close together or quite different?* When thinking about improvements, prompt students to think about equipment to help eliminate sources of error, such as a camera to time how long a seed takes to drop if it falls quickly.

Homework

Students complete an evaluation of their investigation to decide whether or not they have collected valid data.

3.6 Reach: Inheriting characteristics

Theme: Genetics and inheritance

Sub-theme: Inheritance and DNA structure

Biology NC links

- heredity as the process by which genetic information is transmitted from one generation to the next (KS3)
- single gene inheritance and single gene crosses with dominant and recessive phenotypes (KS4)

Learning objective

- Describe the difference between dominant and recessive alleles.

Year 7 and 8 learning objectives to be built on

- Describe how characteristics are inherited
- Give the definitions of DNA, chromosome, and gene

Key Stage 4 link

- Dominant and recessive alleles.

Tier 2 vocabulary	Tier 3 vocabulary
expressed, inheritance	allele, dominant allele, gene, recessive allele

Answers to SB in-text questions

A Blonde and brown **B** Two **C** Dominant: brown eyes, freckles; recessive: blue eyes, receding chin

Lesson plan

Reactivate knowledge

1 Where is genetic material stored?
2 How is genetic material passed from parents to their offspring?
3 What is a gene?

Lesson resources

Activity: *Inheriting eye patches* (student sheet, support sheet, teacher and technician sheet)

Answers: 1 Nucleus **2** Through the egg (mother) and sperm (father) **3** A section of DNA that contains the information for a characteristic

Trigger interest

Write the following keywords on the board: mother, father, egg, sperm, offspring, fertilisation. Ask students to draw a diagram to link these key words to illustrate the concept of inheritance.

Exposition of main content

Use the photo on the Lesson presentation of children with different-coloured eyes to explain that we have two genes for each characteristic: one that we inherit from our mother, and one that we inherit from our father. These genes may be the same or different. Different forms of the same gene are known as alleles. In this example, there is an allele that causes brown eyes and one that causes blue eyes. (In reality, most characteristics are caused by a combination of multiple genes working together. However, a simplified model of inheritance is sufficient at this stage.)

Introduce the concept of dominant alleles. Illustrate that you only need one copy of this allele for it to be expressed. Use Figure 2 in the Student Book to show the inheritance of brown eyes. Then compare this to recessive alleles, using Figure 3, showing the inheritance of blue eye colour.

Finally, show students Figure 4 showing a range of dominant and recessive characteristics. Explain that some characteristics are more common in the population as they are caused by dominant alleles. (Students will need to refer to this diagram again when they answer the questions on the activity sheet.)

Main activity

Activity: Inheriting eye patches (25 mins) Students classify dominant and recessive characteristics in humans. They then answer questions about how eye patches are inherited in dogs.

Review and reflect

Provide students with mini-whiteboards and marker pens. State a human characteristic. Students write down on their mini-whiteboards whether the named characteristic is caused by a dominant or recessive allele and hold up their answers.

Homework

Ask students to research information on cystic fibrosis and Huntington's disease. In both cases, students should include symptoms of the disease and whether the disease is dominant or recessive.

3.7 Reach: Asexual reproduction and cloning

Theme: Reproduction

Sub-themes: Asexual reproduction, Plant reproduction

Biology NC links

- reproduction in humans (as an example of a mammal), including the structure and function of the male and female reproductive systems, menstrual cycle (without details of hormones), gametes, fertilisation, gestation and birth, to include the effect of maternal lifestyle on the foetus through the placenta (KS3)
- reproduction in plants, including flower structure, wind and insect pollination, fertilisation, seed and fruit formation and dispersal, including quantitative investigation of some dispersal mechanisms (KS3)
- genetic variation in populations of a species (KS4)
- the uses of modern biotechnology including gene technology; some of the practical and ethical considerations of modern biotechnology (KS4)

Working scientifically NC link

- use appropriate techniques, apparatus, and materials during fieldwork and laboratory work, paying attention to health and safety

Learning objective

- Describe how clones are produced through the process of asexual reproduction.

Year 7 and 8 learning objectives to be built on

- Describe the process of fertilisation
- Describe the process of fertilisation in plants

Key Stage 4 link

- Asexual reproduction.

Tier 2 vocabulary	Tier 3 vocabulary
clone, gene pool, identical, unethical, variation	asexual reproduction, *cutting*

Answers to SB in-text questions

A Sexual reproduction requires two parents, but only one parent is needed for asexual reproduction **B** Sexual reproduction **C** It is a simple/cheap technique, and new plants are produced quickly

Lesson plan

Reactivate knowledge

1 What is fertilisation?

2 How many parents are needed for sexual reproduction?

3 What is binary fission?

Answers: 1 When the egg and sperm nuclei join together **2** Two **3** When a cell divides into two to form two new identical cells

Lesson resources

Practical: *Plant cuttings* (student sheet, support sheet, teacher and technician sheet)

Trigger interest

Show students the image of bacterial cells dividing in Figure 1 from the Student Book. Ask: *What do you think this is?* When they have correctly identified what the picture is, ask students to explain how this differs from animals reproducing. Use the discussion to come up with the key features of asexual reproduction: only one parent is needed, no mixing of genetic material occurs, an identical copy of the parent (a clone) is produced.

Exposition of main content

Show students the image of the spider plant from Figure 2 in the Student Book. Explain that, as well as reproducing sexually, many species of plants also reproduce asexually; for example, by using plantlets, tubers, or bulbs. Where possible, use real specimens to illustrate these concepts.

Show students Figure 3 in the Student Book of a rosemary cutting being planted. Explain that this technique, commonly used by gardeners, is a form of asexual reproduction. Students will perform this process in the practical. Discuss the advantages of using this technique. Then explain the main disadvantage for all organisms that reproduce asexually: their gene pool is reduced, increasing the chance of a new disease or environmental challenge negatively affecting the population.

Briefly introduce some examples of animal cloning: cell cloning, tissue culture, and cloning embryos. Explain that, although these techniques may help with disease treatment and prevention from extinction, many people think the methods are unethical.

Main practical

Practical: Plant cuttings (30 mins) Students work in small groups to take cuttings from suitable plants. Then they answer questions about asexual reproduction and cloning.

Review and reflect

Students complete a table to compare the processes of sexual and asexual reproduction, including the advantages and disadvantages of each technique.

Homework

Students produce a graphic organiser to summarise the key features of asexual reproduction, including the advantages and disadvantages of growing plants from cuttings, and at least one example of animal cloning.

B3 Chapter 3: Fertilisation and implantation

Introduction to checkpoint intervention

This checkpoint intervention provides suggestions for a lesson to follow up the B3 3 *Fertilisation and implantation* selection checkpoint assessment. Depending on the outcomes of the assessment, these suggestions could move a class towards secure, to target specific learning outcomes, or to consolidate knowledge for students achieving secure outcomes. Use students' outcomes from the checkpoint assessment to address general misconceptions from the content covered. Be prepared to re-cover content as required; three key concepts from the chapter are covered in more detail below.

Checkpoint secure learning outcomes

- Compare the process of sexual reproduction in plants and animals.
- Evaluate the use of some different methods of contraception.
- Interpret diagrams showing the main stages in the menstrual cycle.
- Explain the steps taken to produce valid data in a seed dispersal investigation.
- Describe the difference between dominant and recessive alleles.
- Describe how clones are produced through the process of asexual reproduction.

Student reflection

Encourage students to reflect on whether there were any checkpoint questions they found difficult or straightforward, as well as their preparation for the checkpoint assessment, including how they revised, the time spent on revision, and what areas they could improve on. For more strategies, see the Metacognition chapter or the Metacognition in Key Stage 3 Science guide on Kerboodle.

Intervention activity: Compare the process of sexual reproduction in plants and animals

When answering 'compare' questions, students have a tendency to focus on the differences between two structures or processes and they forget to mention the similarities. The focus of this activity is on comparing the process of sexual reproduction in plants and animals. However, you can use the same activity for tackling any comparison question.

Comparing sexual reproduction

Working as a whole class, students produce two bullet-point lists: the key features of animal sexual reproduction (such as the names of the gametes, where they are produced, where fertilisation takes place) and the key features of plant sexual reproduction. They then use a Venn diagram to identify the similarities and differences between the two processes. Using the diagram as a stimulus, students compare the process of sexual reproduction in both organisms, which could be written as an exam-style answer or described verbally to their partner. For support, students could refer to the information in B3 3.2 *Sexual reproduction*.

Intervention activity: Evaluate the use of some different methods of contraception

'Evaluate' is one of the less common types of exam-style question. As students do not tend to practise this skill very often, time should be spent on explaining how to tackle this type of question. When evaluating, it is important that students write a balanced answer, providing some positive and some negative factors about a method, process, or technique – in this case, about different forms of contraception.

Evaluating different forms of contraception

Students produce a graphic organiser summarising a range of forms of contraception. For each technique, they must include at least one benefit and one disadvantage. The task could be split across small groups, so each group looks at two methods and then shares their information to produce one overall organiser. Explain to students that with this information to hand, they could produce a summary to highlight the benefits and drawbacks of each method of contraception and then draw a short conclusion for each one. For support, students could refer to the information in B3 3.3 *Preventing pregnancy*.

Intervention activity: Explain the steps taken to produce valid data in a seed dispersal investigation

Students often find it difficult to remember the difference between the terms 'accurate', 'precise', 'reproducible', 'repeatable', and 'valid' data; they often use the terms interchangeably. This activity aims to address this issue. Although this activity is based around the seed dispersal investigation, you can use it in any scientific investigation context.

Collecting valid data

Provide students with the key terms: accurate, precise, reproducible, repeatable, and valid, and a description of each. Students start by matching the description to the key term. The focus of this activity is then on valid data (you could alter the focus when using other investigations as a context). To collect valid data, a scientist should control all variables in an investigation apart from the independent variable. Ask students to recall their seed dispersal investigations. Explain that they could be asked to investigate how the mass of the seed affects the distance travelled away from the point of release (the position of the parent plant). Ask students to name:

- the independent variable
- the dependent variable
- the variables that would need to be controlled.

You can extend students' thinking by asking them to explain how each of the control variables would be kept constant. Discuss with students that only if the control variables have been kept constant can data be considered to be valid and hence a definite conclusion formed. For support, students could refer to the information in B3 3.5 *Investigating seed dispersal mechanisms*.

B3 Chapter 4: Variation and natural selection

Introduction to chapter

In this chapter, students will discover why many characteristics are affected by both the environment and inheritance before looking at the different types of adaptations animals have that increase their chances of survival and reproduction. They will investigate organism distribution along a transect and study how antibiotic-resistant bacteria evolve as a result of natural selection. Finally, they will look at the techniques scientists are using to maintain biodiversity, along with the abiotic and biotic factors that affect an organism living in an ecosystem.

Core concepts

- Variation and natural selection

Prerequisite knowledge

- Heredity as the process by which genetic information is transmitted from one generation to the next
- Differences between species
- Changes in the environment which may leave individuals within a species, and some entire species, less well adapted to compete successfully and reproduce, which in turn may lead to extinction
- The variation between species and between individuals of the same species meaning some organisms compete more successfully, which can drive natural selection
- The importance of maintaining biodiversity and the use of gene banks to preserve hereditary material
- The interdependence of organisms in an ecosystem, including food webs and insect pollinated crops
- How organisms affect, and are affected by, their environment, including the accumulation of toxic materials

GCSE links ↗

- The importance of biodiversity
- Positive and negative human interactions with ecosystems
- Some abiotic and biotic factors that affect communities; the importance of interactions between organisms in a community

Key learning misconceptions

- Animals choose to adapt to their environment.
- Adult animals do not change their appearance.
- Camouflaged animals are brown or green.
- Only predators can run fast.
- Plants are prey organisms.
- Evolution happens in individuals rather than populations over time.
- Characteristics are all caused by genetics.
- When organisms are no longer found in one area of the world, they have become extinct.
- Species are organisms that can reproduce, for example, a donkey and a horse are the same species.
- 'Survival of the fittest' relates to strength and dominance between individual organisms.
- Evolution and natural selection are the same thing.
- Human effects on the environment are always negative.

Teaching preparation

Key learning misconception

Animals choose to adapt to their environment Confusion arises from the everyday use of the phrase 'people adapt to changes', causing many students to believe that organisms choose to adapt to their environment. Reinforce the idea that organisms do not adapt quickly but that adaptation of a species is a slow, chance-based change over time, where useful adaptations give an advantage so that organisms with these adaptations survive and reproduce.

Key learning misconception

Evolution and natural selection are the same thing Ensure that students are clear on the distinction between evolution and natural selection. Evolution is a gradual change in the inherited traits of a population over many generations. Natural selection is the mechanism that drives evolution – members of a population best suited to their environment have the best chance of surviving to pass on their genes. Students can think of natural selection in terms of survival of the fittest.

Key literacy skill

Prefixes When helping students to learn the difference between abiotic and biotic, explain that the prefix a- means 'without' or 'not'. Therefore, abiotic means without life – hence, abiotic factors are non-living factors. Other examples include amorphous (without form) or atypical (not typical). Can students think of any others?

Key learning misconception

Human effects on the environment are always negative When asked to describe human effects on the environment, most students will start by listing many negative effects, for example, water pollution or global warming. Challenge students to state one positive effect for every negative effect they give, such as setting up nature reserves or replanting hedgerows.

4.1 Reactivate: Variation and natural selection

Themes: Biodiversity and human interactions with the environment, Ecosystems and environment, Genetics and inheritance, Variation and evolution

Sub-themes: Adaptation, Biodiversity, Conservation, Ecosystems, Evolution, extinction, and the fossil record, Inheritance and DNA structure, Interdependence and competition

Biology NC links

- heredity as the process by which genetic information is transmitted from one generation to the next
- the variation between species and between individuals of the same species meaning some organisms compete more successfully, which can drive natural selection
- the importance of maintaining biodiversity and the use of gene banks to preserve hereditary material
- how organisms affect, and are affected by, their environment, including the accumulation of toxic materials

Learning objective
Recall and organise key knowledge about competition and adaptation.

Year 8 learning objectives to revise

- Describe what is meant by the interdependence of organisms
- Explain the resources that plants and animals compete for
- Describe how organisms adapt to environmental changes
- Use examples to describe the difference between inherited and environmental variation
- Describe how new species evolve through the process of natural selection
- Describe how gene banks can be used to prevent the extinction of a species

Pinchpoint focus

The process of evolution by natural selection

Tier 2 vocabulary	Tier 3 vocabulary
characteristic, competition, evolution, extinct, species, variation	adaptation, fossil, *natural selection*

Trigger interest

Show students an image of a tiger. Ask students to use the image and their own knowledge to explain how it is adapted to survive and reproduce.

Show students an image of kangaroos boxing. Ask them to suggest what the kangaroos could be fighting over and what else animals compete for. Students should then make a list of the factors that plants compete for.

Show students an image of a fossil. Ask them to state what it is and what it provides evidence of. Then ask students to make a list of the factors that could cause an organism to become extinct.

Recall

Provide students with a list of characteristics. Ask students to sort them into those that show environmental variation, inherited variation, or both. They should give reasons for their answers.

Working in pairs, ask students to produce a flow diagram to show the main steps in the evolution of a species. For additional support, supply key phrases for students to include such as 'survival of the fittest', 'genes', and 'advantageous characteristics'.

Main activity

Pinchpoint: Evolution (20–25 mins)

All life exists because it evolves to fit the environmental conditions. Those organisms that adapt best will pass on those characteristics to their offspring. When learning about evolution, students often find it hard to distinguish between what happens to a single organism and to the species as a whole.

Students answer the pinchpoint question and complete the follow-up activity with the same letter. They then complete the *What I need to remember* activity. Information about each incorrect answer's misconceptions can be found on the teacher sheet. You may want to divide the class into groups, depending on the letter answer chosen, to support each other when completing the follow-up activity.

Provide students with images of two cartoon characters. They draw what the offspring may look like, explaining which features are caused by inherited characteristics, environmental characteristics, or both.

Review and reflect

Students are shown the pinchpoint question again and asked to reflect on whether they think they chose the correct letter and if they would change their answer now. If they think they got it wrong, ask them to think about why they chose that answer.

Reveal the correct answer and review with students why it is correct and the other options are not. Guidance on this can be found on the teacher sheet. Students could be asked to pick an incorrect answer and write an explanation to a fictional student who chose it, explaining why it cannot be correct.

Provide students with characteristics to classify as affected by the environment, genes, or both. They arrange their answers into a Venn diagram.

Homework

Students fill in the incomplete knowledge organiser.

4.2 Apply: Variation

Themes: Genetics and inheritance, Variation and evolution

Sub-themes: Inheritance and DNA structure, Variation

Biology NC links

- heredity as the process by which genetic information is transmitted from one generation to the next
- differences between species

Learning objective

- Explain why many characteristics are affected both by the environment and through inheritance.

Year 7 and 8 learning objectives to be applied

- Give the definition of variation
- Use examples to describe the difference between inherited and environmental variation

Application

- Differences between identical and non-identical twins.

Tier 2 vocabulary	Tier 3 vocabulary
characteristic, identical, variation	environmental variation, gene, inherited variation

Answers to SB in-text questions

A Environmental variation **B** Light is needed for photosynthesis to occur to produce food to help the plant grow, while water transports dissolved sugars and other nutrients through the plant, and helps the plant to stand upright **C** 8.2

Lesson plan

Reactivate knowledge

1 What is variation?

2 How is genetic variation passed on?

3 What is a gene?

Lesson resources

Activity: *Twin studies* (student sheet, support sheet, teacher and technician sheet)

Answers: 1 Differences in characteristics within a species **2** Through the egg and sperm/ovule and pollen **3** A section of DNA that codes for a characteristic

Trigger interest

Show students the image of the identical twins in Figure 1 of the Student Book. Ask: *Can you spot any differences? What is likely to have caused these differences?* The differences students may observe, such as a moustache, are results of environmental variation. Explain that identical twins form when a fertilised egg splits. This results in twins who have exactly the same genes. They show no genetic variation (also known as inherited variation). Explain that non-identical twins are formed from separate eggs developing in the uterus at the same time. They share many characteristics. However, the combination of genes they inherit from their parents is not identical, so they can be (for example) different sexes.

Ask students to list characteristics that are a result of genetic variation only. Examples include eye colour and blood group.

Discuss examples of environmental variation using the image of hydrangea flowers in Figure 2 of the Student Book. Examples include your accent and body piercings.

Exposition of main content

Show students the image of sunflowers from the Lesson presentation. Discuss what affects the height to which a sunflower grows. Students should determine that its final height results from a combination of both its genes and access to environmental factors (in particular, light and water). Discuss a range of factors that are affected by both environmental and genetic variation, such as the height and mass of humans, and the yield of fruit trees.

Main activity

Activity: Twin studies (30 mins) Students look at the characteristics of a pair of identical twins. They determine which characteristics are caused by genetics, the environment, or both. They then analyse data from a scientific study on twins.

Review and reflect

Students produce a Venn diagram to sort a number of characteristics into those caused by environmental variation, genetic variation, or both.

Homework

Ask students to answer the exam-style question:

Explain why the mass of corn harvested from a field is a result of both environmental and inherited variation. **(6 marks)**

There are tips on the Lesson presentation to scaffold their answer, if required.

4.3 Apply: More adaptations

Themes: Ecosystems and environment **Sub-theme:** Adaptation

Biology NC link

- changes in the environment which may leave individuals within a species, and some entire species, less well adapted to compete successfully and reproduce, which in turn may lead to extinction

Learning objective

- Describe some structural, behavioural, and functional adaptations of animals.

Year 7 and 8 learning objectives to be applied

- Describe how organisms are adapted to their environment
- Describe how organisms adapt to environmental changes

Application

- Introduce terms structural, behavioural, and functional adaptations.

Tier 2 vocabulary	Tier 3 vocabulary
camouflage, insulation	adaptation, behavioural adaptation, functional adaptation, structural adaptation

Answers to SB in-text questions

A For example: thick coat, layers of fat, small surface area to volume ratio **B** To improve its chances of reproducing **C** To defend themselves against predators

Lesson plan

Reactivate knowledge

1 What is an adaptation?

2 Give some adaptations of a cactus.

3 What is the difference between a predator and a prey organism?

Lesson resources

Activity: *Types of adaptation* (information sheet, student sheet, support sheet, teacher and technician sheet)

Answers: 1 Characteristic that helps an organism to survive **2** For example, spikes, stores water in its stem, thick waterproof layer **3** Predators eat prey organisms

Trigger interest

Show students the image of the polar bear in Figure 1 of the Student Book. Ask them to list its adaptations and give reasons for each adaptation. When sharing students' ideas, encourage the use of scientific terminology such as 'camouflage', 'insulation', and 'surface area to volume ratio'. Discuss that animals don't only have adaptations to help them survive in the conditions in which they live. They also have adaptations that help them to get the food and mates they need to survive and reproduce.

Exposition of main content

Using the examples in the Student Book, introduce the three main types of adaptation: structural, behavioural, and functional.

Begin with structural adaptations, because these are the adaptations with which students will be most familiar. Use the image of the camouflaged cuttlefish in Figure 2 of the Student Book as stimulus.

Next, introduce behavioural adaptations. Use the image of the peacock in Figure 3 of the Student Book. Most students will already be familiar with some species hibernating during the winter, and birds migrating to find food. Many behavioural adaptations are learned, usually through copying.

Finally, introduce functional adaptations. Use the image of the camel in Figure 4 of the Student Book. Functional adaptations are likely to be a new concept. Explain that these are internal processes and cannot be learned by the organism.

Main activity

Activity: Types of adaptation (30 mins) Students complete a card sort activity to categorise adaptations into structural, behavioural, and functional adaptations. They then research an animal of their choice and produce a poster to display its adaptations, categorising each one and giving reasons for the characteristic. Some suggestions for suitable organisms are given in the accompanying resources.

Review and reflect

Provide students with mini-whiteboards and marker pens. Describe an adaptation of an organism, such as 'squirrels store nuts to eat in the winter'. Students write down whether the given adaptation is behavioural, functional, or structural and hold up their answers.

Homework

Students produce a graphic organiser to summarise the three main groups of adaptations: structural, functional, and behavioural, including a range of examples.

4.4 Apply: Investigating competition

Theme: Ecosystems and environment

Sub-theme: Interdependence and competition

Biology NC link

- the variation between species and between individuals of the same species, meaning some organisms compete more successfully, which can drive natural selection

Working scientifically NC links

- select, plan and carry out the most appropriate types of scientific enquiries to test predictions, including identifying independent, dependent, and control variables
- use appropriate techniques, apparatus, and materials during fieldwork and laboratory work, paying attention to health and safety
- make and record observations and measurements using a range of methods for different investigations
- apply sampling techniques

Learning objective

- Describe how to investigate the distribution of plants using a transect.

Year 7 and 8 learning objective to be applied

- Explain the resources that plants and animals compete for

Application

- Investigate how light affects the distribution of plants with distance.

Tier 2 vocabulary	Tier 3 vocabulary
competition, resources, sampling	quadrat, transect

Answers to SB in-text questions

A Plants make their own food by the process of photosynthesis **B** Type and number of plants

Lesson plan

Reactivate knowledge

1 What is competition?

2 Name some factors animals compete for.

3 Name some factors plants compete for.

Lesson resources

Practical: *Investigating plant distribution along a transect* (student sheet, support sheet, teacher and technician sheet)

Answers: 1 Competing with other organisms for resources **2** For example, mates, food, space (territory) **3** For example, light, water, nutrients

Trigger interest

Show students the image of the forest floor in Figure 1 from the Student Book. Ask them to list the factors that the plants are competing for. Ensure that students are aware that plants do not compete for food, as they photosynthesise. However, they do compete for minerals which are needed for healthy growth. Ask students to highlight the factors in their list that animals also compete for, and add any others such as mates.

Exposition of main content

Ask students to discuss how they could investigate the distribution of plants in an area, then share ideas. You could show the image of the quadrat in Figure 3 of the Student Book, to promote thinking. In your discussion, ensure that students know *why* scientists take samples. Ensure students are confident with how to use a quadrat to sample accurately. For example, they need to decide whether to include partly covered plants in their measurements, or not.

Use the image of students using a quadrat in Figure 4 of the Student Book to explain how a transect can be used to study how the distribution of organisms changes as you move across an area. As a class, discuss the method (provided in the Student Book) you could use to investigate how the distribution of plants changes with increasing distance from a hedgerow. In each case, encourage students to think about the factors that the plants may be competing for.

Main activity

Practical: Investigating plant distribution along a transect (35–40 mins) Students plan and carry out an investigation to explore plant distribution along a transect of their choice. They then analyse their findings. This is a fairly long practical, so students may need to complete the analysis for homework.

Review and reflect

Students describe their investigation to another pair and explain their findings.

Homework

Students design a method to investigate the distribution of plant species with a (suitable) given distance from the tide line. This should include a risk assessment and discussion about how the method differs from the method used in the main activity. For instance, if the distance used is 100 m from tide line up the beach, students should realise that a quadrat every 10 m would be more suitable than every 1 m.

4.5 Apply: Natural selection and antibiotic-resistant bacteria

Theme: Variation and evolution

Sub-theme: Evolution, extinction, and the fossil record

Biology NC link

- the variation between species and between individuals of the same species, meaning some organisms compete more successfully, which can drive natural selection

Learning objective

- Explain how antibiotic-resistant bacteria evolve as a result of natural selection.

Year 7 and 8 learning objectives to be applied

- Describe the process of natural selection
- Describe how new species evolve through the process of natural selection

Application

- Antibiotic-resistant bacteria.

Tier 2 vocabulary	Tier 3 vocabulary
evolution, species, sterile, variation	antibiotic, antiseptic, disinfectant, mutation, natural selection

Answers to SB in-text questions

A Animals: food, water, space to live, mates; Plants: light, water, nutrients **B** A mutation could enable bacteria to survive short exposure to high temperatures or starvation **C** Disinfectants are strong chemicals and will damage the skin and area around a cut

Lesson plan

Reactivate knowledge

1 What is evolution?

2 What is meant by 'survival of the fittest'?

3 What is an antibiotic?

Lesson resources

Activity: *Antibiotic resistance* (information sheet, student sheet, teacher and technician sheet)

Answers: 1 Process by which a new species develops as a result of natural selection **2** The best-adapted organisms survive and reproduce **3** Drug that kills bacteria

Trigger interest

Ask students to work in pairs to produce a flow diagram to describe the main steps of natural selection. When sharing students' ideas, ensure that students refer to the fact that the best-adapted organisms survive *and* reproduce in order that the advantageous versions of genes (alleles) are passed on.

Exposition of main content

Using the image of MRSA in Figure 1 of the Student Book, introduce the concept of antibiotic-resistant bacteria. Ask students to share what they know about 'superbugs'.

Explain that all genetic variation occurs as a result of mutations – changes in DNA structure. Explain that bacteria reproduce rapidly and that, when their DNA is copied, mutations can occur. This normally results in the bacteria dying. However, occasionally it can be advantageous, such as the ability to not be killed by an antibiotic. Use the diagram in Figure 2 of the Student Book to explain how this can lead to an antibiotic-resistant population of bacteria.

Using students' knowledge of the Covid-19 pandemic as a starting point, discuss steps that can be taken to prevent the spread of microorganisms. This is a good time to remind students that antibiotics can only be used to treat bacterial infections (so could not be used against coronavirus). Expect students to suggest a range of methods to prevent the spread of disease, such as increased hand washing, wearing masks, and isolation. Conclude the discussion by sharing the importance of taking a full course of antibiotics.

Main activity

Activity: Antibiotic resistance (30 mins) Students work in small groups to rearrange a pack of cards (copies of information sheets, either pre-cut or cut up by the students) to explain how antibiotic resistance can evolve in a species of bacteria, using the example of MRSA. They then use this information to answer questions.

Review and reflect

Show students the definitions of key terms. Ask them to write down the key term being described and hold it up on a mini-whiteboard.

Homework

Students produce a graphic organiser to summarise the range of techniques used to maintain biodiversity. They should state the advantages and disadvantages of any techniques and, where possible, give examples of where each technique is being used.

4.6 Reach: Maintaining biodiversity

Theme: Biodiversity and human interactions with the environment

Sub-themes: Biodiversity, Conservation

Biology NC links

- the importance of maintaining biodiversity and the use of gene banks to preserve hereditary material (KS3)
- the importance of biodiversity (KS4)
- positive and negative human interactions with ecosystems (KS4)

Working scientifically NC links

- interpret observations and data, including identifying patterns and using observations, measurements and data to draw conclusions

Learning objective

- Describe some techniques for maintaining biodiversity.

Year 7 and 8 learning objectives to be built on

- Describe what is meant by the interdependence of organisms
- Describe some factors that may lead to extinction
- Describe how gene banks can be used to prevent the extinction of a species

Key Stage 4 link

- The importance of biodiversity.

Tier 2 vocabulary	Tier 3 vocabulary
advantage, biodiversity, conservation, disadvantage, ecosystem, extinction, prevent, programme, stable	breeding, seed bank

Answers to SB in-text questions

A A rainforest because of high amounts of sunlight and precipitation, along with variations in elevation, leads to greater biological diversity **B** Trampling can adversely affect natural habitats, so preventing it can help to protect habitats **C** An endangered species is seriously at risk of dying out; extinct species no longer exist in the world

Lesson plan

Reactivate knowledge

1 What is extinction?

2 Give some reasons why organisms may become extinct.

3 What is biodiversity?

Answers: 1 No living individuals of a species left anywhere in the world **2** For example, habitat destruction/new disease/new predators/environmental change **3** Range of organisms living in an area

Lesson resources

Activity: *Project Tiger* (information sheet, student sheet, support sheet, teacher and technician sheet)

Trigger interest

Show students the image of the rhino in Figure 1 of the Student Book alongside the images of a woolly mammoth, dodo, and dinosaur from the Lesson presentation. Ask: *Can you spot the difference between them?* (You may wish to add other examples, such as other dinosaurs, to help them find the connection.) The rhino is endangered while the mammoth, dodo, and dinosaur extinct. Ask students to list the factors that can lead to organisms becoming endangered or extinct.

Exposition of main content

Show students the images of a grass field and a forest from the Lesson presentation. Discuss which has the higher biodiversity. Ask students to discuss, in pairs, which is the most stable ecosystem, then share their ideas with the class. Encourage students to talk about the interdependence of organisms. Students should come to the conclusion that, in general, the greater the biodiversity, the more stable the ecosystem.

Ask students to share their ideas of how we can maintain (and in some cases increase) biodiversity. Produce a class list on the board. Group ideas into the broad headings of 'conservation', 'breeding programmes', and 'seed banks'. Most ideas will be forms of conservation (there are lots of examples listed in the Student Book). Where possible, ask students to give examples of projects where the techniques are being used. Discuss the advantages and disadvantages of each technique.

Main activity

Activity: Project Tiger (30 mins) Students read a case study about how tigers have become endangered, and a government initiative in India trying to save them. They then use the information provided and their own knowledge to answer questions.

Review and reflect

Describe a population decrease scenario, such as a decline in the numbers of a species of butterfly in a specific habitat. Students state the most appropriate techniques which could be used to prevent the loss of the species, such as replanting of wild flowers, designated pathways around the edge of protected land, or a ban on the use of pesticides.

Homework

Students produce a graphic organiser to summarise the range of techniques used to maintain biodiversity. They should state the advantages and disadvantages of any techniques and, where possible, give examples of where each technique is being used.

4.7 Reach: Ecosystems – biotic and abiotic factors

Theme: Ecosystems and environment

Sub-themes: Ecosystems, Interdependence and competition

Biology NC links

- the interdependence of organisms in an ecosystem, including food webs and insect pollinated crops (KS3)
- how organisms affect, and are affected by, their environment, including the accumulation of toxic materials (KS3)
- some abiotic and biotic factors which affect communities; the importance of interactions between organisms in a community (KS4)

Working scientifically NC links

- use appropriate techniques, apparatus, and materials during fieldwork and laboratory work, paying attention to health and safety
- make and record observations and measurements using a range of methods

Learning objective

- Use examples to describe the difference between biotic and abiotic factors.

Year 7 and 8 learning objectives to be built on

- Describe what is meant by the interdependence of organisms
- Explain the resources that plants and animals compete for

Key Stage 4 link

- Some abiotic and biotic factors which affect communities; the importance of interactions between organisms in a community.

Tier 2 vocabulary	Tier 3 vocabulary
competitor, ecosystem, predator	abiotic factor, biotic factor

Answers to SB in-text questions

A For example: mates, water **B** For example: soil, pH, wind, humidity

Lesson plan

Reactivate knowledge

1 Name some factors plants compete for.
2 Name some factors animals compete for.
3 What is an adaptation?

Lesson resources

Practical: *Abiotic factors* (student sheet, support sheet, teacher and technician sheet)

Answers: 1 Light, water, space, minerals **2** Food, water, space, mates **3** A characteristic that enables an organism to survive

Trigger interest

Look at the image of the birds migrating in Figure 1 of the Student Book. Ask: *What do you think the birds are doing and why?* Use this to introduce the concept that organisms are affected by both living factors (in this case, a lack of food) and non-living factors in their environment.

Exposition of main content

Introduce the terms *abiotic factors* and *biotic factors.* Remind students that 'bio' refers to 'living', and the prefix 'a-' means 'not', to help them to remember the difference.

Start by looking at biotic factors. What factors can students think of? Most will come up with a food source and the presence of predators. They may need prompting to think of other competitors. Discuss factors that will lead to both an increase (such as a bumper crop) and a decrease in a population size.

Then discuss abiotic factors. Ask students to list as many factors as they can think of. Share ideas to produce a class list. For each example, probe students to think how this factor would affect an organism. For example, plants grow faster in warmer temperatures because photosynthesis occurs faster. There are other examples in the Student Book that can be discussed.

Main activity

Practical: Abiotic factors (40 mins) Students compare the environment in different areas by measuring a range of abiotic factors. They relate this information to the types of organism present in the environment.

Review and reflect

Provide students with mini-whiteboards and marker pens. State a factor. Students then write down whether the given factor is biotic or abiotic on their mini-whiteboard and show their answers.

Homework

Students sketch a diagram of an organism living in a particular habitat. They then label the abiotic and biotic factors present. You may wish to talk through an example, such as a frog in a pond. Students should draw a frog surrounded by pondweed and insects. The insects are a biotic factor, while the pH of the water, the oxygen concentration, and so on are all abiotic factors.

Introduction to checkpoint intervention

This checkpoint intervention provides suggestions for a lesson to follow up the B3 4 *Variation and natural selection* checkpoint assessment. Depending on the outcomes of the assessment, these suggestions could move a class towards secure, to target specific learning outcomes, or to consolidate knowledge for students achieving secure outcomes. Use students' outcomes from the checkpoint assessment to address general misconceptions from the content covered. Be prepared to re-cover content as required; three key concepts from the chapter are covered in more detail below.

Checkpoint secure learning outcomes

- Explain why many characteristics are affected both by the environment and through inheritance.
- Describe some structural, behavioural, and functional adaptations of animals.
- Describe how to investigate the distribution of plants using a transect.
- Explain how antibiotic-resistant bacteria evolve as a result of natural selection.
- Describe some techniques for maintaining biodiversity.
- Use examples to describe the difference between biotic and abiotic factors.

Student reflection

Encourage students to reflect on whether there were any checkpoint questions they found difficult or straightforward, as well as their preparation for the checkpoint assessment, including how they revised, the time spent on revision, and what areas they could improve on. For more strategies, see the Metacognition chapter or the Metacognition in Key Stage 3 Science guide on Kerboodle.

Intervention activity: Describe some structural, behavioural, and functional adaptations of animals

Students generally find it easy to name structural adaptations of animals as these can often be seen. They have also discussed these types of adaptations since Key Stage 2. Difficulties lie with determining the difference between a behavioural and a functional adaptation.

Identifying behavioural and functional adaptations

When thinking about behavioural and functional adaptations, teach students to ask themselves whether the organism could have learned how to do the process. For example, gorillas can learn from others how to use tools. In this case, the adaptation is behavioural. A process that occurs internally, such as the production of venom, is a functional adaptation. Provide students with a number of behavioural and functional adaptations to categorise. They must also give reasons for their classification. For support, students could refer to the information in B3 4.3 *More adaptations*.

Intervention activity: Describe how to investigate the distribution of plants using a transect

Longer-answer exam-style questions in science can often require a student to recall a practical procedure. This requires writing in a coherent sequence, as well as drawing on subject knowledge for the particular context of the question. It should also include a list of the equipment used for the experiment and, where relevant, any calculations that would need to be performed on the data collected.

Using transects

Ask students to write a method for this investigation as a series of bullet points or numbered steps, rather than writing in continuous prose. Students may benefit from being given a series of prompts to help structure their response, to ensure they include all relevant information. For example: equipment list, control variables, how to use a quadrat to collect data, start position, how many positions data will be collected from. For support, students could refer to the information in B3 4.4 *Investigating competition*.

Intervention activity: Explain how antibiotic-resistant bacteria evolve as a result of natural selection

When describing the main steps in the development of a species, encourage students to structure their answer using bullet points or a flow diagram. Once students are confident with drawing a generic flow diagram of this process, they should then find it easy to adapt this to the context of the question being asked – in this case, antibiotic-resistant bacteria.

Main steps in the development of antibiotic-resistant bacteria

Ask students to work in pairs to produce a flow diagram of the main steps in the development of a species. Provide students with key terms that they need to include in their diagram, such as 'variation', 'mutation', and 'genes'. They should then add further detail so it relates directly to the development of antibiotic-resistant bacteria. Students should then swap their diagram with another pair and add any detail they have missed. For support, students could refer to the information in B3 4.5 *Natural selection and antibiotic-resistant bacteria*.

Students should then answer the exam-style question:

Explain how antibiotic-resistant bacteria evolve as a result of natural selection. **(6 marks)**

3 Chemistry

Introduction to unit

This unit consolidates fundamental concepts from earlier books – states of matter, atomic structure, the Periodic Table, chemical reactions, energy changes, and displacement reactions – and offers students practice in vital skills such as using chemical formulae and writing chemical equations. These ideas are extended into important GCSE topics, such as bonding, electronic structure, and relative mass. There is also a significant focus on broader scientific themes like designing safe experiments, models and their limitations, and the history of science.

Chemistry NC links

- The particulate nature of matter
- Atoms, elements, and compounds
- Chemical reactions
- Energetics
- The Periodic Table
- Materials

Working scientifically links

Each lesson lists the relevant Working scientifically National Curriculum links at the top of the spread.

For further details, please refer to the *Oxford Smart* Curriculum for Science document.

Learning journey

Year 7	Year 8	This unit	Key Stage 4 topics
Particles and their behaviour	The Periodic Table	Particle model and state change	Atoms, bonding and moles
Elements, atoms, and compounds	Separation techniques	Atoms and the Periodic Table	Chemical reactions and energy changes
Reactions	Metals and other materials	Chemical changes	Rates, equilibrium, and organic chemistry
Acids and alkalis	The Earth	Useful chemical reactions	Analysis and the Earth's resources

Chemistry and you

Ask students to name a metal, a medicine, a food, and a type of rock. Stress that despite their differing appearances, these materials are all made from atoms from just over 100 elements. Explain that scientists are developing new nanomaterials (materials containing nanoparticles 0.1 mm–0.001 mm in size) with unique properties, offering applications across multiple industries. These materials have these properties because of the arrangement and motion of their nanoparticles.

Ask students what they know about artificial intelligence. Explain that cutting edge programs help chemists design molecules with specific properties and predict the structures of proteins making up all living things. Modern research now combines theoretical, experimental, and computational work, requiring a wide range of expertise. Studying science not only offers all these skills, but also an investigative way of thinking and a curiosity about the world that will be beneficial in many careers and throughout life.

Chemistry and the world

Ask students what they think are the major problems facing scientists today and explain chemistry's role in solving them. Designing efficient catalytic converters and greener energy sources will help to tackle the climate crisis by reducing the emissions of harmful gases. Materials scientists are investigating biodegradable plastics and polymers that do not pollute the environment. Pharmaceutical companies are researching new medicines and production methods, offering novel solutions for tackling diseases.

Highlight that it is more important than ever for students to possess knowledge of chemistry to stay healthy and informed. To help us stay safe and healthy, scientists put new medicines and procedures through rigorous clinical trials and safety checks. Unproven diets, supplements, and procedures should be treated with caution, particularly regarding long-term effects.

Big questions

What's important about chemical reactions? Chemical reactions are important in every aspect of our lives: making life-saving drugs, for transport, producing food, and for every process in every living thing! Remind students that most materials that they encounter on a day-to-day basis are made possible by chemical reactions. Chemical reactions also transfer energy, which we can use to heat our bodies and power our batteries.

What makes substances useful? A substance's properties make them useful to us in our everyday lives, as well as in industries and medicine. Remind students that both the chemical and physical properties of a substance will determine how it interacts and reacts with other substances, and what the substance itself could be used for. It is also worth talking to students about how the state of a substance will change its properties.

Where can chemistry take you? Explain that the skills gained when studying chemistry are just as important as the knowledge learned. Being able to observe, analyse, and interpret a situation are essential in all jobs and in wider life. Aside from skills, an education in chemistry prepares students for a wide variety of careers and tackling decisions in everyday life, such as understanding the warnings on a cleaning product.

C3 Chapter 1: Particle model and state change

Introduction to chapter

The understanding of the arrangement and movement of particles is fundamental to explaining the properties that substances have.

This chapter covers the particle model and state changes that students will have encountered before. The chapter begins with the particle model and then moves onto substances, states of matter, energy changes during state changes, and finally limitations of the model. This chapters also covers sublimation – often a forgotten about state change, as well as foams and aerosols.

This topic will not only be important for GCSE Chemistry, but GCSE Physics as well, when students go on to study temperature, density, state changes, and gas pressure.

Core concepts

- Particle model
- State change

Prerequisite knowledge

- The properties of the different states of matter (solid, liquid, and gas) in terms of the particle model, including gas pressure
- Changes of state in terms of the particle model

GCSE links

- The shapes of molecules (groups of atoms bonded together) and the way giant structures are arranged is of great importance in terms of the way they behave
- Changes of state in terms of particle kinetics, energy transfers, and the relative strength of chemical bonds and intermolecular forces
- Bulk properties of materials related to bonding and intermolecular forces
- Relating models of arrangements and motions of the molecules in solid, liquid, and gas phases to their densities
- Melting, evaporation, and sublimation as reversible changes
- Calculating energy changes involved in heating, using specific heat capacity; and those involved in changes of state, using specific latent heat
- Link between pressure and temperature of a gas at constant volume, related to the motion of its particles (qualitative)

Key learning misconceptions

- Particles in liquids are an intermediate distance apart compared to solids and gases.
- Particles are stationary in solids.
- Particles only exert forces on each other by bumping into each other.
- There is air in between particles.
- The properties of a substance are the same as the properties of the particles that make up that substance.

Teaching preparation

Teaching strategy

Go slowly to begin with Students will have encountered particle theory before in previous study, but this topic is often littered with misconceptions and therefore it is best to take it slowly and proceed with caution. Once students have a clear understanding of the particles in solids, liquids, and gases, they can then move onto state changes and energy transfers when changes of state take place.

Key maths skill

Using ratios in chemical formulae Ratios are used a lot in chemistry, so it is essential that students are comfortable with what the numbers mean and how to manipulate them. It may help to link ratios to situations where students encounter ratios in everyday contexts, such as baking or currency. It is also worth emphasising that students will have studied ratios in maths lessons, and the methods used are the same, just in a different context.

Progression

Sublimation and deposition These state changes are often forgotten about, as most substances go via the liquid state. Two common substances that sublimate are iodine, which has a very impressive purple vapour, and carbon dioxide, known as dry ice in its solid form. Both make for stunning demonstrations, but be careful: if the demonstration is too exciting, students will recall the colours and the smoke and not the science that they are being taught. If you are doing these as demos, think carefully about when you show them to the class.

Key learning misconception

Solids, liquids, and gases Students often struggle with the gas state the most and have a very vague idea about it. Perhaps this stems from not being able to touch gases in the way that we touch solids or liquids (of course we are 'touching' the air all the time, but we often forget this). It might be advantageous to introduce the solid and liquid states first, and then use these to introduce the gas state. This might help with misunderstandings about the spacing of particles (and nothing between them) and their movement.

Key learning misconception

Moving particles Perhaps because diagrams are static, students often forget that particles are constantly in motion (even at absolute zero – they will cover this in GCSE physics). Particles in air at room temperature have an average speed of about 500 m/s! Use of animations and videos will help students appreciate their constant motion.

Modelling

The particle model: The particle model is exactly that: a model. It is not perfect but is a very good way to visualise and understand matter and its properties. Hopefully, students will have discussed scientific models before, for example, conceptualising electric circuits and making models of wind-pollinated seeds. It is important to discuss the nature and limitations of this model. One simple inaccuracy of the particle model is that molecules are not spheres, and it may intrigue your students to know that water molecules are actually V-shaped.

1.1 Reactivate: Particle model and state change

Themes: Atomic structure and the particle model

Sub-theme: Elements, Mixtures, Particles, Solutions, States of matter

Chemistry NC links

- the properties of the different states of matter (solid, liquid and gas) in terms of the particle model, including gas pressure
- changes of state in terms of the particle model.
- the concept of a pure substance

Working scientifically NC links

- understand that scientific methods and theories develop as earlier explanations are modified to take account of new evidence and ideas, together with the importance of publishing results and peer review
- use appropriate techniques, apparatus, and materials during fieldwork and laboratory work, paying attention to health and safety

Learning objective

- Recall and apply key knowledge about particles and state changes.

Year 7 learning objectives to be revised

- Describe the arrangement, separation, and movement of the particles in the three states of matter.

Year 8 learning objectives to be revised

- Use the particle model to explain how evaporation works

Pinchpoint focus

Particles in a gas

Tier 2 vocabulary	Tier 3 vocabulary
aerosol, foam, gas, model, particle, random, separation, substance, theory	Brownian motion, endothermic, exothermic, particle, sublimation

Trigger interest

Show the students a photograph of an iceberg in the sea. Ask: *What states of matter can you see? What properties do these states of matter have?*

Heat a beaker full of crushed ice until the water starts boiling. Ask students to explain what is happening.

Recall

Sketch the arrangements of particles in solids, liquids, and gases. Then, students can describe the motion of particles in these states.

Provide students with the melting and boiling points of different elements, for example, the alkali metals. Ask them what state they would be in at 50 °C and 250 °C.

Main activity

Pinchpoint: Particle model and state change (20–25 mins)

In a gas, the particles move around randomly and the particles are apart from each other. The particles have enough energy to overcome the intermolecular forces between them.

Students answer the pinchpoint question and complete the follow-up activity with the same letter. They then complete the *What I need to remember* activity. Information about each incorrect answer's misconceptions can be found on the teacher sheet. You may want to divide the class into groups, depending on the letter answer chosen, to support each other when completing the follow-up activity.

Students can produce a cooling curve for stearic acid or tert-butanol and plot a graph of the results.

Review and reflect

Students are shown the pinchpoint question again and asked to reflect on whether they chose the correct letter and if they would change their answer now. If they think they got it wrong, ask them to think about why they chose that answer.

Reveal the correct answer. Review with students why it is correct and the other options are not. Guidance can be found on the teacher sheet. Students could be asked to pick an incorrect answer and write an explanation to a fictional student who chose it, explaining why it cannot be correct.

Students can produce a poster summarising the particles in the three states of matter and the state changes between.

Homework

Students fill in the incomplete knowledge organiser.

1.2 Apply: The particle model

Theme: Atomic structure and the particle model

Chemistry NC link

- the properties of the different states of matter (solid, liquid and gas) in terms of the particle model, including gas pressure

Sub-theme: Particles

Working scientifically NC links

- understand that scientific methods and theories develop as earlier explanations are modified to take account of new evidence and ideas, together with the importance of publishing results and peer review
- make and record observations and measurements using a range of methods for different investigations; and evaluate the reliability of methods and suggest possible improvements
- present reasoned explanations, including explaining data in relation to predictions and hypotheses

Learning objectives

- Explain how Brownian motion and dilution are evidence for the particle model.
- Explain how a scientific theory changed over time.

Year 7 and 8 learning objectives to be applied

- Use the particle model to explain the properties of a substance in its three states.

Application

- Explain in detail how ideas about Brownian motion developed over time.

Tier 2 vocabulary	Tier 3 vocabulary
gas, liquid, particle, solid, solution	Brownian motion

SB in-text question answers

A Pollen moving, jiggling about, and turning in water **B** The pollen and small rocks were both moving about randomly and without stopping **C** Random motion of particles due to collisions with other particles **D** The light pink one on the right

Lesson plan

Reactivate knowledge

1. What is the particle model?
2. How do particles move in the liquid state, and how are they arranged?
3. What is a solution?

Lesson resources

Activity: *Demonstration of Brownian motion* (student sheet, support sheet, teacher and technician sheet)

Answers: 1 An idea that describes and explains the movement and arrangement of the particles that make up a substance, and that can be used to make predictions **2** Movement – random, from place to place, sliding over each other; arrangement – random, touching their neighbours **3** A mixture that forms when a substance dissolves in a liquid

Ensure students are confident with the arrangement, movement, and separation of particles in the solid, liquid, and gas states.

Trigger interest

Show students the image of a pink fairy flower on the Lesson presentation and ask: *What role did it play in giving evidence for particles? How is it involved in the story of their discovery?*

Exposition of main content

Ask students to imagine they are watching a football match. What would they think if all of the players became invisible? Would they think the ball had come alive and was moving by itself? *Think pair share* ideas.

Introducing Brownian motion Introduce Scottish biologist Robert Brown and his 1820s experiment with particles in pollen grains. *Think pair share* ideas about why the particles in the pollen were moving.

What did Brown do next? Ask students to suggest what Brown could have done to determine if the pollen moved because it was alive. Explain that he repeated the experiment with pollen from dead flowers and with tiny pieces of other materials. Ask students to suggest what he saw from these experiments. Each time, he observed the non-stop, random motion.

How is Brownian motion evidence for the particle model? Explain that, by the 1880s, scientists had suggested that liquids and gases consisted of invisible, always moving particles. Ask students to suggest how Brown's observations support this theory. Explain that this non-stop, random motion of particles in liquid or gas states is now called Brownian motion.

Main activity

Activity: Demonstration of Brownian motion (25 mins) If you have access to a smoke cell and projector, show students the random motion of smoke particles as they collide with invisible air particles. Or, if you have access to a projectable microscope, show the random motion of tiny polystyrene spheres suspended in a liquid. Alternatively, play the Institute of Physics online video showing small polystyrene spheres colliding with invisible water molecules. Ask students to order the statements to show how the ideas for Brownian motion developed over time.

Modelling Brownian motion Create a model using a speaker and table tennis balls (following the Institute of Physics online video), or discuss the invisible football players model and link the model to the students' visualisations of Brownian motion from the demonstration. Address any misconceptions, such as: The particles that can be seen moving are atoms/molecules, or: The particles are alive/moving by themselves.

Ask students to write a summary explaining why Brownian motion is evidence for the particle model.

Review and reflect

Demonstrate adding potassium manganate to water. *Think pair share* the answer to: *How is this evidence for the particle model?* Explain that particles of the solid potassium manganate (the pink solute) break off and spread apart in the water (the solvent) to produce the pink solution. Show students three different concentrations of a coloured solution, such as potassium manganate. The more particles of the pink solute, the denser they are and the darker the solution is. This shows that both the solute and solvent are made from particles.

Homework

Ask students to create a flow map or knowledge organiser summarising how the ideas about Brownian motion developed over time.

1.3 Apply: Substances

Theme: Atomic structure and the particle model

Chemistry NC links

- differences between atoms, elements and compounds
- the concept of a pure substance
- mixtures, including dissolving

Sub-themes: Elements, Mixtures, Solutions

Working scientifically NC links

- present observations and data using appropriate methods, including tables and graphs
- present reasoned explanations, including explaining data in relation to predictions and hypotheses

Learning objectives

- Explain why something is, or is not, a substance.
- Present data in a bar chart and use it to draw conclusions.

Year 7 and 8 learning objectives to be applied

- Use the particle model to explain why different substances have different properties.
- Give the definitions of an atom and an element (in terms of its atoms).
- Give the chemical formula of a compound, given a labelled molecular diagram.
- Compare mixtures and compounds.

Application

- Analysing the fat content in milk and milk alternatives.

Tier 2 vocabulary	Tier 3 vocabulary
bar chart, compound, conclusion, data, element, molecule, substance	

SB in-text question answers

A For example – compounds: water, carbon dioxide, sugar/sucrose; element – nitrogen
B One **C** It is a mixture of substances

Lesson plan

Reactivate knowledge

1 What is a substance?
2 What is an element?
3 What is a molecule?

Lesson resources

Activity: *Identifying substances* (student sheet, support sheet, teacher and technician sheet)

Answers: 1 A single type of material, with identical particles **2** A substance that cannot be broken down into other substances and that is made up of one type of atom only **3** A group of two or more atoms, strongly joined together

Trigger interest

Ask students: *Do you like cola? Do you know what is in a bottle of cola?* Show the images of the substances in cola and discuss what the students know about them.

Exposition of main content

Elements, molecules, and compounds Divide the class into three groups and give students mini-whiteboards. Each group should recall the definitions of either an element, a molecule, or a compound, and give examples. Share and discuss these as a class, addressing any misconceptions. You could ask students to copy these definitions and some examples into their books.

What is a substance? Recall the definition of a substance in chemistry as something that is made up of only one type of material. All elements and compounds are substances. Discuss the differences between substances and mixtures in chemistry, using the example of a bottle of cola; the cola itself is a mixture, but it is made up of substances such as the compounds carbon dioxide, water, and glucose (sugar), and the element nitrogen. Highlight that the term 'substance' is often used in everyday life to mean something different, such as drugs, but in chemistry it has a specific meaning.

Main activity

Activity: Identifying substances (25 mins) Introduce milk as a mixture of substances. Ask students to suggest some of the substances in milk (water, lactose (a sugar), different fats, some proteins, and some enzymes). Discuss why we would describe milk as a mixture and why water and sugar are substances. Introduce the activity and display the hypothesis on the board. Students determine the correct method of displaying the information, a bar chart (for discontinuous (discrete) data), which they draw, giving it a title and labelling the axes. Students then evaluate the scientist's hypothesis, by writing a short paragraph using data from the table, considering evidence for and against the hypothesis. They then complete the questions, including explaining whether something is a substance or a mixture.

Review and reflect

Students evaluate the answer to the question given on the board. Discuss the strengths and weaknesses of the answer as a class and suggest improvements. Strengths: the answer 'ice' is correct, and it gives the formula of water. It would be given 1 mark. Improvements: The word 'element' is used incorrectly, which loses the second mark; it should say 'type of particle' to be correct. The answer could have given extra information, for example, named ice as water, or said water is a compound; it could also have included a description of it being the same all the way through.

Homework

Ask students to find three materials at home and write a sentence for each material explaining whether it is a substance or a mixture.

1.4 Apply: States of matter

Theme: Atomic structure and the particle model

Chemistry NC link

- the properties of the different states of matter (solid, liquid and gas) in terms of the particle model, including gas pressure

Sub-theme: States of matter

Working scientifically NC links

- present reasoned explanations, including explaining data in relation to predictions and hypotheses
- interpret observations and data, including identifying patterns and using observations, measurements and data to draw conclusions

Learning objectives
• Describe the movement, arrangement, and separation of the particles in foams and aerosols. • Interpret secondary data presented in a bar chart.

Year 7 and 8 learning objectives to be applied	Application
• Describe the arrangement, separation, and movement of particles in the three states of matter.	• Interpret respiratory virus data.

Tier 2 vocabulary	Tier 3 vocabulary
aerosol, foam, gas, liquid, particle, solid	state of matter

SB in-text question answers
A Quickly and randomly from place to place in all directions **B** Gas **C** Gas **D** J

Lesson plan

Reactivate knowledge

1. What are the three states of matter?
2. What is Brownian motion?
3. What is secondary data?

Lesson resources
Activity: *Different states* (student sheet, support sheet, teacher and technician sheet)

Answers: 1 Solid, liquid, and gas **2** The random motion of suspended pieces of solid that occurs as a result of collisions with the randomly moving particles of liquid or gas that the solid is suspended in **3** Data that have been collected by someone else

Trigger interest

Ask students to discuss why foam is used in recording studios (to reduce background noise when recording, to stop sounds reverberating off the walls, for increased soundproofing), before sharing as a class.

Exposition of main content

Introducing foams Ask students to *think pair share* which states of matter can be found in soundproofing foam. Elicit that soundproofing foam is made of materials in the gas and solid states. Describe the structure of soundproofing foam and ask students to describe the movement of particles in the gas state and in the solid state. Elicit squirty cream and firefighting foam as other examples of foam. Ask students to suggest what states of matter the substances are in. Highlight that all foams contain matter in the gas state.

Introducing aerosols Introduce aerosols as another material that contains matter in two different states. Give some examples of aerosols (fog, deodorants, perfumes, smoke, and so on). Ask students to suggest which states of matter are present in these. Highlight that all aerosols contain matter in the gas state.

Main activity

Activity: Different states (20–30 mins) Students interpret data from graphs, then answer an exam-style question.

Interpreting data Inform students that when a person infected with a respiratory virus (such as COVID-19) breathes out, they breathe out tiny particles of virus, either on their own or in little liquid droplets, which mix with the air in an aerosol. Discuss how this may cause others to become ill. Show the graph of the number of droplets of a respiratory virus different people exhaled. Ask students to discuss and determine who breathed out the most droplets (person J) and the fewest droplets (person F). *Think pair share* the meaning of the terms 'threshold of infection' and 'asymptomatic' (they do not show symptoms of the disease), and discuss what it means to be above and below the threshold. Students then analyse the data in the graph to answer questions and create a table to sort people into different groups.

Comparing states of matter Show students the question: *Compare the movement, arrangement, and separation of particles in a foam mattress.* [6 marks]

As a class, discuss how to complete the 'Knowledge of' planning grid. Discuss students' 'Knowledge of task' (the students' understanding of what the task is asking), 'Knowledge of strategies' (the different ways they can go about completing the task), and their 'Knowledge of self' (how confident they are with the subject knowledge the task requires). Students should then complete their 'Knowledge of' planning grid, and use this to help to answer the exam question.

Review and reflect

Ask students to *think pair share* why soundproofing foam on walls reduces the sound travelling through the walls. Students should consider how sound travels through different states of matter (matter in the solid state is a much better conductor of sound than matter in the gas state – so foam with pockets of gas will reduce the conductance), and compare the level of sound that would be conducted through the walls if they were left bare.

Homework

Ask students to identify and name two foams and two aerosols at home. They should then pick one of each and describe the arrangement and movement of the particles in the two states in them. You could suggest examples to use if needed.

1.5 Apply: Sublimation: a change of state

Theme: Atomic structure and the particle model

Chemistry NC link

- the properties of the different states of matter (solid, liquid and gas) in terms of the particle model, including gas pressure

Sub-theme: States of matter

Working scientifically NC links

- make and record observations and measurements using a range of methods for different investigations; and evaluate the reliability of methods and suggest possible improvements
- interpret observations and data, including identifying patterns and using observations, measurements and data to draw conclusions

Learning objectives

- Use the particle model to explain applications of sublimation.
- Interpret secondary data in a line graph and use it to draw conclusions.

Year 7 and 8 learning objectives to be applied

- Describe how the arrangement, separation, and movement of the particles change when a substance evaporates, condenses, and sublimes.

Application

- Interpret sublimation data in air fresheners.

Tier 2 vocabulary	Tier 3 vocabulary
boiling, condense, evaporate, evaporation, vapour	*deposition*, sublimation, subliming

SB in-text question answers

A The change of state from solid to gas without forming a liquid **B** Particles leave from the surface of the solid **C** Some of the carbon dioxide sublimes

Lesson plan

Reactivate knowledge

1 Name the states of matter before and after melting.
2 In evaporation, from where do particles leave the liquid?
3 What type of graph is useful for plotting continuous data?

Lesson resources

Practical: *Sublimation of air freshener* (student sheet, support sheet, teacher and technician sheet)

Answers: 1 Before – solid, after – liquid **2** The surface **3** Line graph

Trigger interest

Show students Figure 1 from the Student Book (a snow-covered mountain top) and ask: *On the southern slopes of Mount Everest, the snow disappears without melting. How?*

Ask students to discuss answers in pairs, and write a suggested answer on mini-whiteboards, but do not confirm it; they will come back to this at the end of the lesson.

Exposition of main content

Recalling the states of matter Ask students to recall the three states of matter, and the names for the changes of state. Draw and label a diagram on the board showing the states of matter and changes of state. Highlight sublimation (solid state to gas state) and deposition (gas state to solid state) as the changes of state they might not be so familiar with. Recap these processes, including the arrangement, movement, and separation of particles in the beginning and end states.

Applications of sublimation Ask students to suggest when sublimation is useful. Elicit the use of dry ice as 'fog' in stage shows, and for keeping COVID-19 vaccinations very cold for periods of time without making a mess, the use of iodine sublimation in the revealing of fingerprints in forensics, and the use of sublimation as a separation technique.

Main activity

Practical: Sublimation of air freshener (20–25 mins) Introduce air fresheners (and mothballs) as smelly substances that can undergo sublimation at room temperature. Students observe sublimation of an air freshener and calculate the mass that sublimes over a period of time. They answer questions about the experiment and calculate the rate of sublimation of the air freshener.

Interpreting line graphs While the equipment is cooling, discuss the interpretation of line graphs. Students should fill in their observations and complete the questions. Introduce naphthalene, another substance that sublimes, used in mothballs. Show the graph of the sublimation rate of naphthalene and discuss how to interpret it.

Review and reflect

Return to the question from the beginning of the lesson. Students should evaluate their previous answer and improve on it on the mini-whiteboards. Discuss why it might be important to climate scientists to be aware of, and study, sublimation of snow in the mountains (to monitor the water loss from, and snow coverage on, mountainsides; to monitor the rate of climate change and impact on sea level rise; to monitor the water and mass balance on glaciers).

Homework

Ask students to use the particle model to describe the changes in the movement, arrangement, and separation of particles of naphthalene as it sublimes, and give an advantage and a disadvantage of using naphthalene in mothballs all year round.

1.6 Reach: Energy transfer in changes of state

Theme: Atomic structure and the particle model

Chemistry NC links

- changes of state in terms of the particle model (KS3)
- energy changes on changes of state (qualitative) (KS3)
- changes of state of matter in terms of particle kinetics, energy transfers and the relative strength of chemical bonds and intermolecular forces (KS4)

Sub-themes: Particles, States of matter

Working scientifically NC links

- ask questions and develop a line of enquiry based on observations of the real world, alongside prior knowledge and experience
- present reasoned explanations, including explaining data in relation to predictions and hypotheses

Learning objectives

- Use the particle model to explain energy changes in changes of state.
- Interpret data presented in a table.

Year 7 and 8 learning objectives to be built on

- Describe how the arrangement of particles changes when a substance changes state.

Key Stage 4 link

- Measuring energy changes in chemical reactions

Tier 2 vocabulary	Tier 3 vocabulary
combustion, freeze, melt, reaction, temperature	change of state, endothermic, exothermic

SB in-text question answers

A When a substance melts, energy is transferred *from* the surroundings to the substance **B** Iron has a higher melting point than aluminium, so the furnace does not get hot enough to melt iron **C** Endothermic – two from evaporation, sublimation, melting; exothermic – one from condensing, freezing

Lesson plan

Reactivate knowledge

1 What is an endothermic change?
2 What is an exothermic change?
3 Describe what happens to the particles when a liquid evaporates.

Lesson resources

Practical: *Observing energy transfer in changes of state* (student sheet, support sheet, teacher and technician sheet)

Answers: 1 A chemical reaction or physical change in which energy is transferred from the surroundings to the substances involved in the change **2** A chemical reaction or physical change in which energy is transferred to the surroundings from the substances involved in the change **3** Particles leave the surface of the liquid. In the air, they spread out and move around quickly and randomly from place to place, in all directions

Trigger interest

Show students Figure 1 from the Student Book of people melting aluminium. Explain that they melt aluminium from old, unexploded bombs, and remould it to make spoons. Ask students to discuss why they need to heat aluminium to melt it.

Exposition of main content

Recalling changing states Discuss melting and boiling points occurring at different temperatures, and the need for particles to have enough energy for melting/boiling to occur. Recall the changes in the arrangement and movement of particles as substances melt and boil.

Exothermic and endothermic changes Hand around an activated instant cold pack and an instant heat pack. Ask students to recall which shows an exothermic process (heat pack) and which shows an endothermic process (cold pack). Recall and highlight where energy is transferred from and to. In endothermic processes, thermal energy is transferred from the surroundings into the substances. In exothermic processes, thermal energy is transferred from the substances to the surroundings. Give students tips to help them remember which is which: EXothermic – thermal energy EXits the substance, ENdothermic – thermal energy ENters the substance. Link back to the idea that melting and boiling both require energy and so are endothermic changes. Students should draw a labelled diagram to help them to remember the energy transfers that occur during endothermic and exothermic processes.

Student modelling With a group of about ten students, model particles in the solid state. Ask students to suggest what would happen if you gave them more (heat) energy (they would move around more). Using the student model, show the substance changing state, highlighting how we know it is now in the liquid state.

Main activity

Practical: Observing energy transfer in changes of state (20–25 mins) Students observe and record the temperature change of water in the solid state (ice) melting, as energy is transferred from the surroundings to the substance the ice. They should identify this as an endothermic change. They may also see the water in the air cooling and condensing on the outside of the test tube as it transfers energy from the substance (the water vapour) to the surroundings (the cold test tube). This is an exothermic change but is often harder for students to comprehend. While students wait for the state change to occur, they can finish their diagrams or complete some of the questions using the data in the table.

Review and reflect

Encourage *think pair share* discussion of some metacognitive reflection questions to help students to evaluate their learning. For example: *Are there areas you feel you still need to revise here? Did you understand the practical? What strategies did you employ to answer the questions? Did they work? How could you improve them? What went well in the lesson? What could have gone better?*

Homework

Ask students to write a short paragraph, using the particle model, to explain why puddles drying up is an endothermic process.

1.7 Reach: Limitations of the particle model

Theme: Atomic structure and the particle model

Chemistry NC links

- the properties of the different states of matter (solid, liquid and gas) in terms of the particle model, including gas pressure
- changes of state in terms of the particle model
- a simple (Dalton) atomic model

Sub-theme: Particles

Working scientifically NC links

- ask questions and develop a line of enquiry based on observations of the real world, alongside prior knowledge and experience (KS3)
- make predictions using scientific knowledge and understanding (KS3)
- using a variety of concepts and models to develop scientific explanations and understanding (KS4)

Learning objective
• Evaluate the strengths and weaknesses of the particle model.

Year 7 and 8 learning objectives to be built on
• Use the particle model to explain why different substances have different properties.
• Give the four factors in the particle model that determine the properties of a substance.
• Give the factor in the particle model that explains why different substances have different melting points.

Key Stage 4 link
• Evaluate the strengths and limitations of the particle model.

Tier 2 vocabulary	Tier 3 vocabulary
evaluate, model	

SB in-text question answers
A For example, they both have wheels **B** For example, marbles **C** Copper **D** For example, in the particle model, all particles are spherical; in reality, some particles – for example, water – have different shapes

Lesson plan

Reactivate knowledge

1 What is a model in science?

2 What is the particle model?

3 In the particle model, what shape are particles assumed to be?

Lesson resources
Activity: *Strengths and limitations of models* (student sheet, teacher and technician sheet)

Answers: 1 An idea that explains observations and helps in making predictions **2** An idea that describes and explains the movement and arrangement of the particles that make up a substance, and can be used to make predictions **3** Spherical

Ensure students are confident with the arrangement, movement, and separation of particles in the solid, liquid, and gas states.

Trigger interest

Show students Figures 1 and 2 from the Student Book, the images of the model truck and the real truck, and discuss the similarities and differences between them.

Exposition of main content

What are models and why are they useful? Ask students to consider why we might make models of trucks. How might they be useful? Discuss that we can use a model to explain observations, to make plans, and to make predictions. Ask students to recall the different models we have used in science (for example, the particle model, student models, ball, marble, balloon, coin models, building brick models) and why these are useful. Discuss that often in science we use models to help represent concepts that are abstract or not observable in real life, and use the models to make predictions or explain ideas.

Main activity

Activity: Strengths and limitations of models (25–35 mins) Provide small groups of students with building bricks and assign them objects to make small models of, for example, a table, cup, straw, apple, flower, or car. In their groups they should discuss the strengths and limitations of their models. It may be useful to stop the class part way through this task to share ideas.

The particle model – strengths and limitations Students apply this thinking to the particle model and, as a class, discuss some of the strengths and limitations associated with the particle model. Students complete the table with some of these ideas, and answer the questions.

Review and reflect

Give students mini-whiteboards and ask them to write 'strength' on one side and 'limitation' on the other. Display a statement on the board and ask students to hold up their mini-whiteboard showing if it is a strength or a limitation of the particle model.

Homework

Ask students to write a short summary paragraph evaluating the strengths and limitations of the particle model.

Chapter 1: Particle model and state change

Introduction to checkpoint intervention

This checkpoint intervention provides suggestions for a lesson to follow up the C3 1 *Particle model and state change* checkpoint assessment. Depending on the outcomes of the assessment, these suggestions could move a class towards secure, to target specific learning outcomes, or to consolidate knowledge for students achieving secure outcomes. Use students' outcomes from the checkpoint assessment to address general misconceptions from the content covered. Be prepared to re-cover content as required; three key concepts from the chapter are covered in more detail below.

Checkpoint secure learning outcomes

- Explain how Brownian motion and dilution are evidence for the particle model.
- Explain how a scientific theory changed over time.
- Explain why something is, or is not, a substance.
- Describe the movement, arrangement, and separation of the particles in foams and aerosols.
- Use the particle model to explain applications of sublimation
- Use the particle model to explain energy changes in changes of state.
- Evaluate the strengths and weaknesses of the particle model.

Student reflection

Encourage students to reflect on whether there were any checkpoint questions they found difficult or straightforward, as well as their preparation for the checkpoint assessment, including how they revised, the time spent on revision, and what areas they could improve on. For more strategies, see the Metacognition chapter or the Metacognition in Key Stage 3 Science guide on Kerboodle.

Intervention activity: Explain how Brownian motion and dilution are evidence for the particle model

Brownian motion is evidence of particle collisions. All particles move around randomly because they are being constantly hit by other particles. Larger particles, such as pollen grains, can appear to move in a random manner, causing students to believe that the larger particles are moving around randomly of their own accord. It is important to stress that the pollen grains are in fact jiggling because they are constantly colliding with many smaller particles of water.

You can demonstrate Brownian motion with a large loudspeaker, a signal generator, some ping pong balls, and a partially inflated balloon. Place the ping pong balls onto the speaker and set it to vibrate. The ping pong balls will move around randomly like the particles in a liquid. Then add the balloon. Encourage students to focus only on the balloon and imagine the ping pong balls are invisible.

Intervention activity: Describe the movement, arrangement, and separation of the particles in foams and aerosols

Students are likely to misunderstand that foams and aerosols are in both the solid/liquid and the gas states. If students are struggling, it is worthwhile ensuring that they have a secure understanding of the particle model in solids, liquids, and gases. Foams are trapped pockets of gas in solids or liquids and aerosols are small particles of solids or liquid suspended in a gas.

Having ensured that students have a secure understanding of solids, liquids, and gases, it can be helpful to show them foams and aerosols and see if they can distinguish between them *and* explain why. Examples could include methane bubbles (before igniting them) (foam), a sponge (foam), deodorant (aerosol), mist (aerosol), and ice cream (foam).

Intervention activity: Use the particle model to explain energy changes in changes of state

One of the most surprising facts when learning about the particle model is that the temperature does not change when a pure substance changes state. To truly understand this, students need to first understand that temperature is related to how fast particles are moving (their kinetic energy). When a substance is changing state, the particles are moving so fast they can overcome the intermolecular forces between them. The energy being added is used to increase their potential energy (how far apart they are) and not their kinetic energy (how fast they are moving).

Model an answer for the following question:

Describe what happens to the particles in a liquid when it evaporates. **(4 marks)**

Many students will describe the arrangement and motion of the particles. Then, encourage students to re-read their (or another student's) answer and make improvements using the terms: (kinetic) energy, movement, intermolecular forces.

C3 Chapter 2: Atoms and the Periodic Table

Introduction to chapter

Students will have encountered the Periodic Table before and to many students, it is the iconic image of chemistry in their minds. This chapter starts with the development of the Periodic Table, then the structures of atoms, metals and non-metals, Group 2 elements, and compounds, before concluding with electron configuration and bonding.

By introducing atomic structure, electron configuration, and types of bonding, students will be in an excellent place to progress to GCSE.

Core concepts

- Periodic Table
- Atoms and elements
- Compound molecules

Prerequisite knowledge

- The simple (Dalton) atomic model
- Differences between atoms, elements, and compounds
- Chemical symbols and formulae for elements and compounds
- The principles underpinning Mendeleev's Periodic Table
- The Periodic Table: periods and groups; metals and non-metals
- How patterns in reactions can be predicted with reference to the Periodic Table
- the properties of metals and non-metals

GCSE links

- A simple model of the atom consisting of the nucleus and electrons, relative atomic mass, electronic charge, and isotopes.
- The number of particles in a given mass of a substance.
- The modern Periodic Table.
- Position of elements in the Periodic Table.
- Properties and trends of elements in the same group.
- Characteristic properties of metals and non-metals.
- Chemical reactivity of elements in the Periodic Table.
- Types of chemical bonding: ionic, covalent, and metallic.
- Bulk properties of metals related to bonding and intermolecular forces.

Key learning misconceptions

- You can see atoms under a microscope.
- Atoms are alive, like cells, because they both have a nucleus.
- Atoms 'want' a full outer shell.
- Bonding between atoms involves gravitational or magnetic forces.
- The properties of a substance are the same as the properties of the particles that make it up.
- Atoms look like the diagrams.

Teaching preparation

Curriculum coherence

Use history Although Mendeleev is often 'the' scientist credited with the Periodic Table, many individuals contributed to its development. This chapter gives an overview, but if you have time and ability with your classes, it might be worthwhile to encourage them to research others who contributed to the Periodic Table. The structure of the atom is covered at GCSE more thoroughly, but, if time permits, students can also research parts of this story as well.

Key maths skill

Interpreting graphs and tables The properties of elements, such as melting points, in a group are often summarised in a table or graph. Although students have had practice in interpreting tables and graphs, they often write short answers that do not fully describe the results, which loses them marks. Encourage students to be as detailed as they can in their answers, referencing data, including outliers, from tables and graphs as evidence for their conclusions.

Teaching strategy

Electronic structure This will be covered again, but clear modelling of how to write electronic structure is important here. It might be tempting to ask students to complete electronic configurations for the first 20 elements in order, but this can lead students to just adding another electron without thinking about why. Presenting them with the first 20 elements in alphabetical order may prove to be helpful as they must stop and think each time. The 'A' elements (Al and Ar) have enough electrons to serve as good 'I do' examples. The 'B' elements (B and Be) are much shorter for 'we do' examples. 'C' elements (C, Ca, and Cl) are a good start for 'you do' examples.

Teaching strategy

Atomic scales Many diagrams do not show atoms to scale, so students struggle to grasp the vastness of an atom. A typical atomic size is 10^{-10} m and a typical nucleus size is 10^{-14} m, so a nucleus is about 10 000 times smaller than an atom. If your classroom/laboratory is 10 m wide, then a 1 mm speck of sticky tack makes a scale approximation of an atom. Present students with a range of objects (e.g., an apple, a football, a button, the eye of a needle) and ask them to name another object of the same size as the nucleus (or atom) if the first object was the size of the atom (or nucleus).

Key literacy skill

Calcium and compounds The most likely place that students will have heard of calcium before is in food, and so they will be surprised discover that calcium is quite a reactive metal. Emphasise to students that when we say 'milk contains a lot of calcium', what we mean is 'milk contains a lot of calcium compounds'. This is a good example for demonstrating the difference between scientific and everyday language.

Teaching strategy

Bonding There are likely to be many difficulties with teaching bonding the first time. Assure students that they will cover this again, so they will have more opportunities to practise. A key take-home message is that covalent bond corresponds to sharing electrons and ionic bond corresponds to transferring electrons. This simplification might help with students' cognitive load while they are still novices.

2.1 Reactivate: Atoms and the Periodic Table

Theme: Atomic structure and the particle model

Sub-themes: Electrons, The Periodic Table

Chemistry NC links

- the simple (Dalton) atomic model
- differences between atoms, elements and compounds
- chemical symbols and formulae for elements and compounds
- chemical reactions as the rearrangement of atoms

Working scientifically NC link

- understand that scientific methods and theories develop as earlier explanations are modified to take account of new evidence and ideas, together with the importance of publishing results and peer review

Learning objective

- Draw electronic structures for atoms

Year 7 learning objectives to be revised

- State the meaning of atom.

Year 8 learning objectives to be revised

Pinchpoint focus

- Writing electron configurations

Tier 2 vocabulary	Tier 3 vocabulary
compound, configuration, element, formulation, mixture	atom, atomic number, covalent bond, electron, ionic bonding, mass number

Trigger interest

Show students pictures of a helium balloon and of lithium metal reacting with water. Ask them to think about why one is very reactive and the other isn't, even though there is only a difference of one electron between them.

Display the Periodic Table on the board. Ask students what the numbers in each box mean.

Recall

Recap the meaning of atomic and mass number, and how these relate to protons, neutrons, and electrons. Students need to understand that for electron configurations, they need the atomic number, which is the lower of the two.

Display the symbol for one element, e.g., aluminium.
Ask students to write down everything they can work out from this information.

Main activity

Pinchpoint: Atoms and the Periodic Table (20–25 mins)

Students need to be able to write and/or draw electron configurations for the first 20 elements of the Periodic Table. To do this, students need to recall that the first shell can only contain two electrons and thereafter each shell can contain eight electrons.

Students answer the pinchpoint question and complete the follow-up activity with the same letter. They then complete the *What I need to remember* activity. Information about each incorrect answer's misconceptions can be found on the teacher sheet. You may want to divide the class into groups, depending on the letter answer chosen, to support each other when completing the follow-up activity.

If students are comfortable with electron configurations, provide them with a list of all twenty elements listed in alphabetical order and ask them to write out electron configurations. Providing them in alphabetical order and not atomic number order ensures that students are not just adding one electron to the previous answer.

Review and reflect

Students are shown the pinchpoint question again and asked to reflect on whether they chose the correct letter and if they would change their answer now. If they think they got it wrong, ask them to think about why they chose that answer.

Reveal the correct answer. Review with students why it is correct and the other options are not. Guidance can be found on the teacher sheet. Students could be asked to pick an incorrect answer and write an explanation to a fictional student who chose it, explaining why it cannot be correct.

Students choose different elements in the Periodic Table, work out the (correct) electron configuration, along with two or three 'wrong' configurations, and then provide explanations as to why these could be mistakes students might make.

Homework

Students fill in the incomplete knowledge organiser.

2.2 Apply: The development of the Periodic Table

Theme: The Periodic Table

Sub-theme: The Periodic Table

Chemistry NC links

- the principles underpinning the Mendeleev Periodic Table
- the Periodic Table: periods and groups; metals and non-metals
- how patterns in reactions can be predicted with reference to the Periodic Table

Working scientifically NC link

- understand that scientific methods and theories develop as earlier explanations are modified to take account of new evidence and ideas, together with the importance of publishing results and peer review

Learning objective

- Describe the contributions of different scientists to the development of the Periodic Table.

Year 7 and 8 learning objectives to be applied

- Explain what is shown in the Periodic Table.

Application

- Describe the contributions of different scientists to the discovery of the Periodic Table.

Tier 2 vocabulary	Tier 3 vocabulary
element	atom, atomic mass, Periodic Table

SB in-text question answers

A Dalton, Cannizzaro, Mendeleev **B** For elements he had predicted should exist, but had not yet been discovered **C** Ida Tacke and Walter Noddack

Lesson plan

Reactivate knowledge

1. What is an element?
2. Approximately how many elements are there?
3. What is the Periodic Table?

Answers: 1 A substance that cannot be broken down into other substances **2** 100 **3** The Periodic Table lists all the elements and groups together elements with similar properties

Lesson resources

Activity: *Finding elements* (information sheet, student sheet, teacher and technician sheet)

Trigger interest

Ask students to describe how they communicate with their friends. Ask them to suggest how scientists used to communicate in the 1800s, and if they think communication is better/easier nowadays.

Exposition of main content

Asking questions Explain that lots of scientists contributed to the modern Periodic Table and ordering its elements. They used different approaches, experiments, and evidence, but all built on each other's ideas and asked questions of each other. Ask students to name the scientist who is first recognised as having left gaps in the Periodic Table for undiscovered elements (Mendeleev). *Think pair share* how the scientists might have reacted when they were told there was information they hadn't worked out yet. Highlight that his opinion was not popular but eventually evidence of the missing elements were found and their properties matched the predictions that Mendeleev made. Explain that once scientists realised there were still undiscovered elements, they asked questions such as, 'How many elements are there?' and, 'Can we use patterns to predict new elements?'

Sharing knowledge Explain that one of the founding principles of the scientific community is the sharing of knowledge, building on existing ideas and models by collaboration and through mutual accountability. Divide students into groups and ask them to compare the methods of communication used by scientists 300 years ago with those used now. Share ideas as a class. Discuss the advantages and disadvantages of each (for example, speed, accessibility, reach, direct communication with individual scientists, and reliability). Link back to the Trigger interest section and previous lessons about fake news and knowing what sources of information can be trusted; discuss if students still think communication of information is (or isn't) better nowadays.

Main activity

Activity: Finding elements (15 mins) Students use the information sheet to complete a table about scientists and their contributions to the Periodic Table.

False/fake news As a class, discuss how students would feel if they saw a social media post about someone discovering a new element or curing a disease. Would they believe it? Why, or why not? Ask them to consider if it were written up as a news article, would they believe it? Why, or why not? Discuss what fake news is. State that, if something is not 100% true, it is not necessarily 'fake news': people can make mistakes, or social media posts can be rumours, satirical, or biased. Ask students to suggest how we can work out whether to trust information; for example, using multiple sources, having a general understanding of the world, and asking: is there evidence? Link back to the scientific community supporting each other with peer reviews, so we are more able to trust information.

Review and reflect

Students complete a plus, minus, interesting (PMI) grid in their books detailing something that went well in the lesson, something they found more complex, and something they found interesting.

Homework

Ask students to compare the modern Periodic Table with Mendeleev's and suggest some differences between them. For instance, Mendeleev's table had gaps for missing elements whereas scientists are confident that the modern Periodic Table has no missing elements; the modern Periodic Table orders elements by atomic number (whereas Mendeleev's was ordered by atomic mass and did not include atomic number); and Mendeleev's table has eight groups, whereas the modern Periodic Table has 18.

2.3 Apply: Inside atoms

Theme: Atomic structure and the particle model

Sub-theme: Atomic structure

Chemistry NC link

- a simple (Dalton) atomic model

Working scientifically NC links

- understand that scientific methods and theories develop as earlier explanations are modified to take account of new evidence and ideas, together with the importance of publishing results and peer review
- use and derive simple equations and carry out appropriate calculations

Learning objectives

- Describe the structure of an atom.
- Define and use models and analogies.

Year 7 and 8 learning objectives to be applied

- Give the definitions of an atom and an element (in terms of its atoms).

Application

- Model of the atom.

Tier 2 vocabulary	Tier 3 vocabulary
analogy	electron, neutron, *nucleus*, proton, sub-atomic particle

SB in-text question answers

A To explain the different physical properties in the solid, liquid, and gas states; to make predictions about changes of state **B** A few particles bounced straight back off the gold foil

Lesson plan

Reactivate knowledge

1 What is the particle model?

2 What is an atom?

3 What is a model in science?

Lesson resources

Activity: *Making atomic models* (student sheet, teacher and technician sheet)

Answers: 1 An idea that describes and explains the movement and arrangement of the particles that make up a substance, and that can be used to make predictions **2** The smallest part of an element that can exist **3** An idea that explains observations and helps in making predictions

Trigger interest

Show students Figure 1 from the Student Book and ask them to describe the shape of the balls. Explain that they are solid spheres, then ask students to give examples of other objects that are solid spheres.

Exposition of main content

The solid sphere model Ask students to recall the particle model, and what it helps to explain. Discuss how the particle model uses the idea that atoms are solid spheres, and this helps us to explain some of the properties of substances in the three states of matter. Introduce analogies as a comparison that helps to explain something. Give the analogy of atoms being like pool balls.

Improving the particle model Describe the development of our understanding of the structure of an atom when three scientists, Rutherford, Geiger, and Marsden, performed the gold foil experiment. Describe the set-up of the gold foil experiment, showing the image. Ask students to predict what they would expect to happen. Describe how the scientists predicted that most positive particles would go straight through, but that a few would change direction slightly. Explain that the results (a few particles bounced straight back off the foil) were amazing and prompted scientists to rethink how the atom was structured.

Main activity

What is in the nucleus? Explain that scientists continued to do experiments and discuss the results and evidence, and created a model. Explain how their model explains why atoms of different elements have different atomic masses, and why atoms of different elements react in similar or different ways in chemical reactions. Explain that the atom is so small that a different scale is used to make the numbers easier to understand and work with (which is why relative masses and charges, not actual masses and charges, are used).

Ask students to suggest what the sub-atomic particles were that scientists discovered. Discuss their names, relative masses, relative charges, and locations. Complete the table on the board with this information. Highlight that protons and electrons have opposite charges but atoms have equal numbers of protons and electrons, so atoms have no overall electrical charge.

Explain that we can calculate the relative atomic masses of elements by adding their numbers of protons and neutrons. That means that most of the atom's mass is in its nucleus and, because the nucleus is so small compared to the size of the atom, most of an atom is actually empty space. Use the analogy of a football stadium (the atom) and a pea (the nucleus of the atom) on the centre spot, to show the approximate size difference between an atom and its nucleus.

Activity: Making atomic models (15–20 mins) In groups, students create a poster showing the structure of an atom. They label the three different sub-atomic particles, and give information about their relative masses, relative charges, and locations. Students should include an explanation as to why atoms have no overall charge.

Students then complete the questions.

Review and reflect

Using targeted teacher-led questioning and mini-whiteboards, ask students to determine the relative mass, relative charge, or location of each of the sub-atomic particles.

Homework

Ask students to create a knowledge organiser highlighting the names, relative masses, relative charges, and locations of the three sub-atomic particles. It should include an explanation as to why atoms have no overall charge, and an analogy about the size of the nucleus compared with the size of an atom.

2.4 Apply: Metals and non-metals in the Periodic Table

Theme: The Periodic Table

Sub-themes: Metals and non-metals, The Periodic Table

Chemistry NC links

- the Periodic Table: periods and groups; metals and non-metals
- the properties of metals and non-metals
- the varying physical and chemical properties of different elements
- how patterns in reactions can be predicted with reference to the Periodic Table

Working scientifically NC links

- understand that scientific methods and theories develop as earlier explanations are modified to take account of new evidence and ideas, together with the importance of publishing results and peer review
- make predictions using scientific knowledge and understanding
- present reasoned explanations, including explaining data in relation to predictions and hypotheses

Learning objectives

- Classify an element as a metal or non-metal, given its properties.
- Describe some contributions of Mendeleev to the discovery of the Periodic Table.

Year 7 and 8 learning objectives to be applied

- Describe the physical properties of typical metal and non-metal elements.

Application

- Classifying elements as metals or non-metals

Tier 2 vocabulary	Tier 3 vocabulary
chemical property, metal, physical property	non-metal

SB in-text question answers

A Li, Na, K, Rb, Cs, Fr **B** He, Ne, Ar, Kr, Xe, Rn
C Argon

Lesson plan

Reactivate knowledge

1 What are protons?

2 What are the vertical columns in the Periodic Table?

3 Why did Mendeleev leave gaps in his Periodic Table?

Lesson resources

Activity: *Classifying elements* (student sheet, teacher and technician sheet)

Answers: 1 Positively charged sub-atomic particles in the nucleus of an atom **2** Groups **3** For elements that he predicted should exist, but that had not been discovered

Trigger interest

Ask students: *What do you remember about the reaction of potassium with water?* Discuss answers.

Discuss which group potassium is in, and why elements are arranged together in groups.

Exposition of main content

The Periodic Table Ask students to recall how the Periodic Table is arranged (by atomic number and in groups and periods) and where to find metals and non-metals in the Periodic Table.

Swapping elements Discuss how Mendeleev tried to arrange elements in atomic mass order (rather than by atomic number – as this hadn't been discovered yet!). However, after some thinking, he went against his decision and swapped them around. Ask students why he did that and highlight that he knew they needed to be arranged in groups of similar elements, even if this seemingly went against atomic mass order. Highlight that Henry Mosely showed this was the correct thing to do once protons had been discovered and atomic number could be used.

Main activity

Demonstrations of typical properties of metals and non-metals Recap the properties of typical metals and non-metals with demonstrations of the conductivity of electricity and heat, shininess density, and malleability. Ask students to predict (or recall) the properties before each demonstration.

Activity: Classifying elements (15 mins) Divide the students into an even number of groups. Give each group a piece of flipchart paper and pens. Assign each group either metals or non-metals; students should write this in the centre of their paper. Give them five minutes to create a mind map, writing as much information about metals or non-metals as they can. This could be physical and chemical properties, uses, where to find them on the Periodic Table, discovery, and so on. Swap the pieces of paper between the groups (so metals go to non-metals, and vice versa). Give the students another five minutes to read through the mind map and add anything they can. As a class, discuss the mind maps, correct and highlight any misconceptions, and add anything that is missing.

Attach the mind maps to the walls or board and then, as a class, classify some elements based on the information on the mind maps.

Students then use the mind maps to create tables showing information about metals and non-metals, and complete the questions.

Review and reflect

'What am I?' Teacher reads out a description of an elements properties and students have to decide whether it is a metal or non-metal. This can be done by writing on mini-whiteboards or students can move to different sides of the room to vote. Ask students to justify their choice.

Homework

Ask students to research one metal and one non-metal element that they have not met in class yet. They should find out at least five different facts, uses or properties of the two elements and produce mini-fact files about them, classifying them as either metal or non-metal and explaining a reason for each.

2.5 Apply: Groups of the Periodic Table

Theme: The Periodic Table

Sub-themes: Metals and non-metals, The Periodic Table

Chemistry NC links

- the varying physical and chemical properties of different elements
- the principles underpinning the Mendeleev Periodic Table
- the Periodic Table: periods and groups; metals and non-metals
- how patterns in reactions can be predicted with reference to the Periodic Table

Working scientifically NC links

- use appropriate techniques, apparatus, and materials during fieldwork and laboratory work, paying attention to health and safety
- present observations and data using appropriate methods, including tables and graphs
- interpret observations and data, including identifying patterns and using observations, measurements and data to draw conclusions

Learning objectives

- Describe patterns in the properties of the Group 2 elements.
- Present and interpret data in bar charts.

Year 7 and 8 learning objectives to be applied

- From data presented in tables or bar charts, describe patterns in the properties of elements in groups or periods.

Application

- Investigating the properties of Group 2 elements.

Tier 2 vocabulary	Tier 3 vocabulary
element, group, hardness, period	

SB in-text question answers

A The hardest element is beryllium, at the top of the group. The elements get less hard/softer from top to bottom of the group. **B** Calcium chloride and hydrogen **C** For example, eggs, broccoli, yoghurt

Lesson plan

Reactivate knowledge

1 What are the columns in the Periodic Table called?

2 What are physical properties?

3 What is melting point?

Lesson resources

Practical: *Comparing the reactivity of Group 2 elements* (student sheet, teacher and technician sheet)

Answers: 1 Groups **2** Properties that you can observe or measure without changing a material **3** The temperature at which a substance changes from the solid to the liquid state

Trigger interest

Show students Figure 1 from the Student book (the James Webb Space Telescope). Ask: *do you know what a space telescope is?* Explain that the mirror segments are made from the metal beryllium, coated with gold. Ask students to suggest the conditions the space telescope might need to endure and, therefore, what properties it would be useful for its constituent materials to have.

Exposition of main content

Why choose beryllium? Ask students to suggest the properties of beryllium if it is used on a space telescope. Explain that it is strong, has low density, and holds its shape well over a range of temperatures. Explain that beryllium is a Group 2 metal and that, like all groups,the elements of Group 2 show trends in properties.

Why are Group 2 elements useful? Ask students to suggest foods that contain Group 2 elements, suggest why calcium is important in our diets, and recall what colour magnesium burns in oxygen. Tell students that barium (greens) and strontium (reds/purples) burn with bright colours. Ask them to suggest how this property makes them useful on Bonfire Night, Diwali, or other festivals.

Main activity

Practical: Comparing the reactivity of Group 2 elements (30 mins) Students add Group 2 metals, magnesium and calcium, to test tubes containing hydrochloric acid, and compare the reactivities. Students then perform the hydrogen pop test on the reaction between magnesium and hydrochloric acid to prove the production of hydrogen. They then complete the questions, determine the correct method of displaying data (a line graph) on the density and atomic radius of the Group 2 elements, and determine if the data prove or disprove a hypothesis.

Review and reflect

Show students the tables of the hardness and density of Group 2 elements. Ask students to *think pair share* to describe the pattern shown in the data. Link back to the Trigger interest section and ask students to use the trends in hardness and density to suggest and justify if any other Group 2 elements might be suitable for use in a space telescope.

Homework

Ask students to research two uses of the Group 2 elements and write one sentence for each use to explain why the properties of the element make it suitable for its use ('fit for purpose').

2.6 Apply: Elements, compounds, and mixtures

Themes: Atomic structure and the particle model, Bonding and properties of materials

Sub-themes: Atomic structure, Chemical formulae, Elements, Mixtures, Particles, Properties of materials

Chemistry NC links

- differences between atoms, elements and compounds
- chemical symbols and formulae for elements and compounds
- mixtures, including dissolving

Working scientifically NC links

- understand and use SI units and IUPAC (International Union of Pure and Applied Chemistry) chemical nomenclature
- present reasoned explanations, including explaining data in relation to predictions and hypotheses

Learning objectives

- Explain how the uses of elements, compounds, and mixtures depend on their properties.
- Use ratios to determine chemical formulae.

Year 7 and 8 learning objectives to be applied

- State the meaning of atom, element, and compound.

Application

- Researching information about formulations.

Tier 2 vocabulary	Tier 3 vocabulary
compound, element, formulation, mixture	

SB in-text question answers

A Strong, does not crack easily, low density
B 1:1 **C** For example, paints, tablets/medicines, toothpastes

Lesson plan

Reactivate knowledge

1 What is an element?
2 What is a compound?
3 What is a mixture?

Lesson resources

Activity: *Why choose that material?* (student sheet, support sheet, teacher and technician sheet)

Answers: 1 A substance that cannot be broken down into other substances / a substance made of one type of atom **2** A substance made up of atoms of two or more elements, with the atoms joined together strongly **3** A substance that contains two or more substances whose particles are not joined together, but just mixed up

Trigger interest

Show students the image of the tooth implant in Figure 1 of the Student Book. Ask: *What substance is this made from? Why might it be made from that material? What are the properties you'd expect?*

Exposition of main content

Why choose titanium? Explain that the tooth implant is made from titanium because it is a strong, low-density metal.

Explain that titanium can make different compounds. Titanium nitride, TiN, has a high melting point and is very hard so it is used for coating the surface of cutting tools and drills. Ask students to determine the ratio of elements in titanium nitride, TiN (Ti:N, 1:1).

Explain that titanium dioxide, TiO_2, another compound of titanium, is bright white, and is in the solid state at room temperature, so it is used in lots of white mixtures such as toothpaste and paint. Ask students to determine the ratio of atoms in titanium dioxide, TiO_2 ($Ti:O_2$, 1:2).

Elements, compounds, and mixtures Ask students to recall the differences and similarities between elements, compounds, and mixtures, and to give examples of each. Complete the table on the board with students' ideas, addressing any misconceptions.

Main activity

Activity: Why choose that material? (20–25 mins)
Ask students what we need to think about when choosing the right material for a use.

Define the term *formulation* as a mixture designed to do something useful. Describe formulations as being made of carefully measured amounts of different substances that each have a specific purpose. Ask students to think of examples of some formulations (paints, toothpastes, medicines, fuels, cleaning products, fertilisers, foods, alloys, vaccines, and so on).

Give students access to research materials, such as books or the Internet. They need to choose a formulation and create an information leaflet, fact file or article about it: its properties, the substances in it, the properties of the substances in it, and its uses.

Students then explain how the properties of two of the substances in their formulation make it suitable for its use.

Review and reflect

Show students the formulae of some substances. Ask them to determine the ratios of the atoms of each element and write them on mini-whiteboards.

Show students the ratios of atoms of each element in some compounds, and ask students to determine the chemical formulae of those compounds.

Homework

Ask students to choose one formulation that they have in their home and describe its function (job). They should then look at the ingredients list and see if they can pick out some components and describe their role in the overall product. For example, maize starch in paracetamol tablets is to bulk out the tablet, sodium metabisulfite is a preserving agent to improve the shelf life, and magnesium stearate is a lubricant to prevent the paracetamol from sticking to the equipment during manufacture of the tablet.

2.7 Reach: Electronic structure

Themes: Atomic structure and the particle model, The Periodic Table

Sub-themes: Atomic structure, Elements, The Group 1 elements, The Periodic Table

Chemistry NC links

- a simple (Dalton) atomic model (KS3)
- a simple model of the atom consisting of the nucleus and electrons, relative atomic mass, electronic charge and isotopes (KS4)
- position of elements in the Periodic Table in relation to their atomic structure and arrangement of outer electrons (KS4)
- properties and trends in properties of elements in the same group (KS4)
- chemical reactivity of elements in relation to their position in the Periodic Table (KS4)

Working scientifically NC links

- understand that scientific methods and theories develop as earlier explanations are modified to take account of new evidence and ideas, together with the importance of publishing results and peer review
- ask questions and develop a line of enquiry based on observations of the real world, alongside prior knowledge and experience
- make predictions using scientific knowledge and understanding

Learning objectives

- Draw the electronic structure of the atoms of the first 18 elements of the Periodic Table.
- Make predictions using scientific knowledge.

Year 7 and 8 learning objectives to be built on

- Give the definitions of an atom and an element (in terms of its atoms).

Key Stage 4 Link

- Make predictions using scientific knowledge.

Tier 2 vocabulary	Tier 3 vocabulary
particle	atom, electron, electronic structure, neutron, proton, *shell*

SB in-text question answers

A 47 **B** 8 **C** 5

Lesson plan

Reactivate knowledge

1 What is a chemical property?

2 Name the three sub-atomic particles.

3 What is the electric charge on an electron?

Answers: 1 A property that describes how substances change in chemical reactions **2** Proton, neutron, electron **3** –1

Lesson resources

Activity: *Modelling electronic structures* (student sheet, teacher and technician sheet)

Trigger interest

Ask students to recall the difference between physical and chemical properties, and to recall the physical properties of the metallic elements zirconium and sodium. Discuss how their physical properties are similar, but their chemical properties are different as they have differing arrangements of electrons.

Exposition of main content

Arranging electrons Recall the three sub-atomic particles found in an atom and their relative charges, relative masses, and locations. Discuss that atoms are neutrally charged, as they have the same number of negatively charged electrons as positively charged protons. Give a few examples, asking students to determine how many electrons exist in atoms of different elements given the number of protons, and vice versa. Prove this using a calculation if needed (for example, +23 –23 = 0).

Show the image of the arrangement of electrons in an atom of sodium, explaining the arrangement of electrons (two in the first shell closest to the nucleus, eight in the second shell, and one in the outer shell). Highlight that it has one electron in its outer shell.

Main activity

Activity: Modelling electronic structures (25 mins) Give pairs of students a mini-whiteboard, pens, and some beads or counters. Show the electron arrangement of sodium on the board and ask students to create a model of the electron configuration of an atom of sodium on their mini-whiteboards. Repeat this with another one of the first 18 elements, first determining how many electrons an atom of the element must have (based on the number of protons it has) and then arranging them with a maximum of two electrons in the first shell and a maximum of eight in the second shell.

Ask students to complete the table with the number and arrangement of electrons in an atom of each of the first 18 elements, using the image on the board to help if needed. Ask students to discuss, in groups, the patterns in the electron arrangements. Highlight that each element in a group has the same number of electrons in their outer shell and each element in a period has electrons in the same number of shells.

Link back to sodium, showing that an atom of sodium has one electron in its outer shell, so it is in Group 1, and that this is the same for all other Group 1 elements. Explain that, although scientists didn't realise it when the Periodic Table was compiled, this is why elements in the same group have similar chemical properties – because they have the same number of electrons in their outer shell.

Remind students why atoms have an overall neutral charge and discuss how we can use this to determine the electron arrangement in an atom of an element. Ask students to add labels and notes to their tables, and then answer the questions.

Review and reflect

Show students the electron configurations of an atom of an element (sulfur), and ask them to determine which group the element must be in, and which element it is. Ask them to determine which elements will have similar chemical properties (Group 6 elements, such as O and Se). Ask students to reflect on why zirconium and sodium have different chemical properties and to write a short explanation in their books.

Homework

Ask students to create a knowledge organiser that highlights all the information we know about how electrons are arranged in an atom of an element and why atoms have an overall neutral charge.

2.8 Reach: Bonding

Theme: Atomic structure and the particle model

Sub-theme: Particles

Chemistry NC links

- a simple model of the atom consisting of the nucleus and electrons, relative atomic mass, electronic charge and isotopes (KS4)
- chemical reactivity of elements in relation to their position in the Periodic Table (KS4)
- types of chemical bonding: ionic, covalent, and metallic (KS4)

Working scientifically NC links

- make predictions using scientific knowledge and understanding
- present reasoned explanations, including explaining data in relation to predictions and hypotheses

Learning objectives
• Define covalent and ionic bonding. • Make predictions using scientific knowledge.

Year 7 and 8 learning objectives to be built on
• Explain whether a molecule diagram shows an element or a compound.

Key Stage 4 Link
• Ionic and covalent bonding

Tier 2 vocabulary	Tier 3 vocabulary
	covalent bond, ion, ionic bonding

SB in-text question answers
A Sodium: 1; chlorine: 7 **B** A shared pair of electrons **C** Sodium and chloride **D** Electrostatic

Lesson plan

Reactivate knowledge

1 Which sub-atomic particles are outside the nucleus?

2 Give two properties of chlorine.

3 Give two properties of sodium.

Lesson resources
Activity: *Modelling bonding* (student sheet, teacher and technician sheet)

Answers: 1 Electrons **2** For example, green, gas at room temperature **3** For example, shiny solid at room temperature, good conductor of electricity and thermal energy

Ensure students are confident with the arrangement, movement, and separation of particles in the solid, liquid, and gas states.

Trigger interest

Ask students if they know what table salt is made of. Explain that sodium and chlorine atoms join together to make sodium chloride, and that sodium, chlorine, and sodium chloride all have different properties. Ask students to recall why the compound sodium chloride has different properties to the elements it is made from, chlorine and sodium. Explain that the particles in sodium and the particles in chlorine are different and are joined together differently to the particles in sodium chloride.

Exposition of main content

Introducing covalent bonding Explain that covalent bonding occurs between non-metal elements. The bond inside a chlorine molecule is a covalent bond and consists of two chlorine atoms sharing a pair of electrons. Ask students to determine how many electrons are in the outer shell, and explain that each chlorine atom contributes one electron to the shared pair to form the covalent bond. Explain that this occurs so that each atom gains a full outer shell of electrons, which makes them stable. Ask students to write the definition of a covalent bond, a shared pair of electrons, in their books.

Introducing ionic bonding Explain that ionic bonding is another type of bonding. It occurs between metal elements and non-metal elements. Explain how the ionic bonding in sodium chloride occurs: one electron moves from the metal element to the non-metal element to form two charged particles called ions. Show the image of the charged ions being formed and explain that these ions are attracted to one another in a regular arrangement, forming a giant ionic structure. Ask students to write the definition of an ionic bond, the electrostatic attraction between positive and negative ions, in their books.

Main activity

Activity: Modelling bonding (20–25 mins) Students model covalent and ionic bonding using beads, pipe cleaners, and mini-whiteboards. They then complete the table by placing the sentences summarising ionic and covalent bonding, before completing the questions.

Covalent bonding As a class, make a model of a hydrogen molecule, H_2, using pipe cleaners and beads to model the covalent bond between hydrogen atoms. Then display the dot-and-cross diagram for a chlorine molecule, discuss the sharing of a single pair of electrons between the chlorine atoms, and make a model of Cl_2 as a class. Using the dot-and-cross diagrams, students then make models of the bonding in HCl and CH_4. In each case, explain how many electrons are shared between each atom in each molecule.

Ionic bonding Students draw two outer electron shells next to one another on their mini-whiteboards, placing seven beads of one colour on the first mini-whiteboard to represent an atom of chlorine, and one bead of another colour on the second mini-whiteboard to represent an atom of sodium. Students can then show the transfer of the sodium outer shell electron to the chlorine atom, forming two ions. This can be repeated by making models of atoms of lithium and fluorine and then forming lithium fluoride.

Review and reflect

Students create a Venn diagram containing statements comparing ionic and covalent bonding (e.g., 'Mostly occurs between metals and non-metals' would be for ionic bonding, but 'Involves shared pairs of electrons' would be covalent bonding). These statements can either be provided for them or made up by the student.

Homework

Ask students to write two sentences summarising their learning on ionic and covalent bonding.

C3 Chapter 2: Atoms and the Periodic Table

Introduction to checkpoint intervention

This checkpoint intervention provides suggestions for a lesson to follow up the C3 2 *Atoms and the Periodic Table* checkpoint assessment. Depending on the outcomes of the assessment, these suggestions could move a class towards secure, to target specific learning outcomes, or to consolidate knowledge for students achieving secure outcomes. Use students' outcomes from the checkpoint assessment to address general misconceptions from the content covered. Be prepared to re-cover content as required; three key concepts from the chapter are covered in more detail below.

Checkpoint secure learning outcomes

- Describe the structure of an atom.
- Describe the contributions of different scientists to the discovery of the Periodic Table.
- Classify an element as a metal or a non-metal, given its properties.
- Describe patterns in the properties of the Group 2 elements.
- Explain how the uses of elements, compounds, and mixtures depend on their properties.
- Draw electronic structures for atoms.
- Define covalent and ionic bonding.
- Use ratios to determine chemical formulae.

Student reflection

Encourage students to reflect on whether there were any checkpoint questions they found difficult or straightforward, as well as their preparation for the checkpoint assessment, including how they revised, the time spent on revision, and what areas they could improve on. For more strategies, see the Metacognition chapter or the Metacognition in Key Stage 3 Science guide on Kerboodle.

Intervention activity: Use ratios to determine chemical formulae

Working out chemical formulae is an essential skill in chemistry. The sooner students gain confidence with this, the better. Starting with compounds made from non-metal elements can be a useful gateway, as often these include information about the numbers of each atom in their name, for example, carbon monoxide (CO).

With students, deduce the number associated with the prefixes: mono-, di-, tri-, tetra-, penta-, and hexa-. Tri-, penta-, and hexa- can be remembered by association with the 2D shapes (triangle, pentagon, hexagon). For tetra-, remind students that a tetrahedron is a regular four-sided 3D shape. Emphasise that, in chemistry, di- is the prefix for two, not bi- (as in bicycle).

The students can write out the formulae of the following compounds: carbon dioxide, nitrogen monoxide, nitrogen dioxide, dinitrogen monoxide, silicon dioxide, sulfur dioxide, sulfur trioxide, phosphorus pentachloride, sulfur hexafluoride, and diphosphorus pentoxide.

Students can then write out the ratio of each atom in each compound. For example, in carbon dioxide, the carbon and oxygen atoms are in a 1 : 2 ratio.

(NB: compounds of metals with non-metals tend not to include numerical prefixes; for example, $CaCl_2$ is known as calcium chloride, not calcium dichloride. Students will learn at GCSE how to deduce formula of compounds from their ions.)

Intervention activity: Define covalent and ionic bonding

Understanding bonding allows students to explain the properties of different chemicals and materials. A covalent bond is a shared pair of electrons. An ionic bond is the electrostatic attraction between oppositely charged ions. The ions are formed by electrons being transferred between atoms. Covalent bonds normally happen between two non-metal elements (e.g., hydrogen and oxygen in water) whereas ionic bonds form between metals and non-metals (e.g., sodium and chlorine in sodium chloride).

Drawing electron configurations out will help students connect non-metals (with nearly full outer shells) and metals (nearly empty outer shells). For metals, it is *easier* to lose electrons than gain or share them. For non-metals, it is *easier* to gain electrons, which can be done by accepting electrons (hence forming a negative ion) or sharing them.

Students can draw out electron configurations of magnesium, sulfur, and hydrogen. They can then work out the types of bonding in a compound of magnesium and sulfur (MgS: ionic), magnesium and hydrogen (MgH_2: ionic), and sulfur and hydrogen (H_2S: covalent). Ensure that students know that if hydrogen is in a compound with a metal, it will be an ionic compound and if it is in a compound with another non-metal, then it will be covalent.

You could replace sulfur with oxygen, and this is often done as an introductory example. Using sulfur makes students think about electrons instead of jumping to the answer because they happen to know that H_2O is covalent!

Intervention activity: Draw electronic structures for atoms

Electronic structure explains the chemical properties of atoms. It is an extremely important skill needed in early their GCSEs. There are two main sources of confusion with regards to writing electron structures. Firstly, students can often use the mass number instead of the atomic number from the Periodic Table. Secondly, students misremember the number of electrons in each shell.

The first hurdle can be overcome by the students having a firm understanding of atomic structure including proton and neutrons. The second hurdle is harder for students to understand at this stage of their academic careers. The reason for there being two electrons in the first shell is far beyond Key Stage 3 or 4 level. It is just a fact students need to learn at this stage. A placeholder answer at this stage is to indicate the first few periods (rows) of the Periodic Table and point out how many elements are in each one: the same number as the electrons in each shell. That said, students should be able to connect the number of electron shells with the period that the element is in e.g. sodium is in period 3 so has 3 shells.

The best way for students to master this is lots of practice repeated over a period of time. As a teacher, you could have a pot containing with the first twenty element (H to Ca). For every lesson after the lesson where you taught electron configurations, pull a random element from the pot and ask students to work out the electron configuration. This could be done on a mini-whiteboards or the back of their exercise books. Such an activity would be short, and so unlikely to interfere with future lesson and easy to embed into a routine. The power of this activity is in its consistency.

C3 Chapter 3: Chemical changes

Introduction to chapter

Chemical reactions are one of the most interesting parts of chemistry and one that students find most enjoyable. Students first need to understand the difference between a chemical and a physical change, reactants and products, and how to write and balance simple chemical equations. Then they look into different types of chemical reactions. After this, students will learn about conservation of mass, combustion, and exothermic and endothermic reactions.

Core concepts

- Word equations
- Conservation of mass

Prerequisite knowledge

- Chemical reactions as the rearrangement of atoms.
- Representing chemical reactions using formulae and using equations.
- Combustion, thermal decomposition, oxidation and displacement reactions.
- Exothermic and endothermic chemical reactions (qualitative).

GCSE links

- Balanced chemical equations, ionic equations, and state symbols.
- Measurement of energy changes in chemical reactions (qualitative).
- Bond breaking, bond making, activation energy, and reaction profiles (qualitative).

Key learning misconceptions

- Chemical reactions are events (e.g., colour changes), rather than the formation of new substances.
- Energy and heat are products of a reaction.
- Changes of state, like the change from water into steam, are chemical reactions as a new substance has been formed.

Teaching preparation

Teaching strategy

Balancing equations This is a very important skill that students will find very tricky to begin with. Consider starting with simple equations of the form $A + B \rightarrow C$ or $A \rightarrow B + C$. It is also helpful to start with reactions where at least one reactant or product is an element, for example, Mg or O_2.

Broader context

Practical work: A word of warning Chemical reactions are often the most memorable parts of chemistry for students and it is likely that they will remember chemistry practical work into their adult lives, even if they don't study chemistry beyond GCSE. Nevertheless, the excitement of practical work can cause cogitative overload (i.e., students focus more on the excitement of the bubbles and bangs rather than the underlying chemistry). When completing practical work with your class, ensure they have a firm understanding at the molecular level to get the most benefit from practical work. Don't forget that, as teachers/ instructors, you can make connections far more easily than you students can.

Curriculum coherence

Exothermic and endothermic Ensure you have a clear understanding of how energy is taught in physics if chemistry and physics are taught separately to your class or in your school.

Key literacy skill

Exothermic and endothermic Consider the etymology of the words *exothermic* and *endothermic*. *Exo-* means 'out of' (think 'exit') and *therm-* means heat. Heat (energy) exiting the system (the chemicals) means that the surroundings will feel warmer. *Endo-* means 'into' (think 'enter'), so heat (energy) entering the system (the chemicals) means that the surroundings will feel colder.

Key learning misconception

Conservation of mass When demonstrating this principle to students, start with reactions that do not involve gases, so it is easier to make clear that no atoms are created, destroyed, or lost. Precipitation reactions are good for this. Then, move onto reactions that involve gases: students can measure the mass of gas created, or calculate it using the difference in mass of the reactants and the solid or liquid products.

Key maths skill

Rearranging and solving equations When solving problems involving the conservation of mass, students are required to rearrange the terms in a linear equation. There are several methods of finding missing masses (e.g., using formal algebra, subtract appropriate values from both sides, moving terms from one side of the equation to the other whilst changing sign): allow students to use their favourite method, depending on their confidence.

3.1 Reactivate: Chemical changes

Themes: Chemical reactions

Chemistry NC links

- chemical reactions as the rearrangement of atoms
- representing chemical reactions using formulae and using equations
- combustion, thermal decomposition, oxidation and displacement reactions
- conservation of mass changes of state and chemical reactions.

Sub-theme: Understanding chemical reactions

Working scientifically NC links

- understand that scientific methods and theories develop as earlier explanations are modified to take account of new evidence and ideas, together with the importance of publishing results and peer review
- use appropriate techniques, apparatus, and materials during fieldwork and laboratory work, paying attention to health and safety

Learning objective

- Apply the idea of conservation of mass.

Year 7 learning objectives to be revised

- Explain why the total mass of reactants is equal to the total mass of products in a chemical reaction.

Year 8 learning objectives to be revised

Pinchpoint focus

- Conservation of mass.

Tier 2 vocabulary	Tier 3 vocabulary
combustion, product, reaction	atom, conservation of mass, endothermic, exothermic, reactant, word equation

Trigger interest

- Ask: *Which of these will have the same mass before and after?*

Before	After
Two pieces of bread, a slice of cheese, and a piece of lettuce	Cheese sandwich
A new candle	The same candle after it has been burning for a few hours
An uninflated balloon	An inflated balloon

Ask students to write down what they understand by the words ‘conservation’ and ‘mass’.

It may be helpful to link the words.

Recall

Explain that the word 'conserve' means 'preserve' or 'keep intact'. There may be local or national conservation projects you could refer to here.

Conservation means keeping the same. So, conservation of mass means keeping the mass the same.

Ask students to calculate the mass of products or reactants in simple addition or decomposition reactions. For example:

$CuCO_3$	$\rightarrow$	CuO	+	CO_2
		2.4 g		1.3 g

Main activity

Pinchpoint: Chemical changes (20–25 mins)

The conservation of mass applies to all chemical reactions. It is always true because in chemical reactions, atoms rearrange themselves and are not created or destroyed. The conservation of mass can be used by chemists to work out the masses of products in a reaction.

Students answer the pinchpoint question and complete the follow-up activity with the same letter. They then complete the *What I need to remember* activity. Information about each incorrect answer's misconceptions can be found on the teacher sheet. You may want to divide the class into groups, depending on the letter answer chosen, to support each other when completing the follow-up activity.

Students complete an exercise where they predict the mass of products using ratios. Question 4 and in-text question C in the Student Book give examples of this.

Review and reflect

Students are shown the pinchpoint question again and asked to reflect on whether they chose the correct letter and if they would change their answer now. If they think they got it wrong, ask them to think about why they chose that answer.

Reveal the correct answer. Review with students why it is correct and the other options are not. Guidance can be found on the teacher sheet. Students could be asked to pick an incorrect answer and write an explanation to a fictional student who chose it, explaining why it cannot be correct.

Ask students why the conservation of mass is true. Lead them to the idea that atoms are rearranged and not created or destroyed. There would be opportunity for discussion of where the conservation of mass appears not to be true when gases are involved in a reaction.

Homework

Students fill in the incomplete knowledge organiser.

3.2 Apply: Chemical and physical changes

Themes: Atomic structure and the particle model, Chemical reactions

Sub-themes: States of matter, Understanding chemical reactions

Chemistry NC links

- the properties of the different states of matter (solid, liquid and gas) in terms of the particle model, including gas pressure
- mixtures, including dissolving
- chemical reactions as the rearrangement of atoms
- energy changes on changes of state (qualitative)

Working scientifically NC links

- make and record observations and measurements using a range of methods for different investigations; and evaluate the reliability of methods and suggest possible improvements
- interpret observations and data, including identifying patterns and using observations, measurements and data to draw conclusions
- present reasoned explanations, including explaining data in relation to predictions and hypotheses

Learning objectives

- Compare physical and chemical changes.
- Interpret observations to make conclusions.

Year 7 and 8 learning objectives to be applied

- Compare chemical reactions to physical changes.

Application

- Using metacognitive strategies to compare chemical and physical changes

Tier 2 vocabulary	Tier 3 vocabulary
compare	chemical reaction, physical change

SB in-text question answers

A Sodium fumarate, carbon dioxide, water **B** 5
C Two from: melting, boiling, evaporating, subliming, freezing, condensing, dissolving, mixing

Lesson plan

Reactivate knowledge

1 What is a physical change?
2 What is a chemical reaction?
3 What is a base?

Lesson resources

Activity: *Comparing chemical and physical changes* (student sheet, teacher and technician sheet)

Answers: 1 A change that does not make a new substance **2** A change in which atoms rearrange and join together differently to make new substances **3** A substance that can react with acids and neutralise them

Trigger interest

Ask students to suggest the ingredients found in cake, and to name the ingredient that helps cake to rise (baking powder). Discuss how it makes cakes rise (baking powder is a mixture of acids and a base. These two substances dissolve in the water in the cake batter, then the two solutions react in an acid-base reaction which releases bubbles of carbon dioxide gas, making the cake rise). Ask students to suggest how we know this is a chemical reaction.

Exposition of main content

Recognising chemical and physical changes
Demonstrate some chemical reactions (e.g., burning a match, rusting nail) and physical changes (e.g., melting an ice cube, dissolving sugar in water). In pairs, students should observe each demo, make a note of their observations, and identify whether it shows a physical change or a chemical reaction on a mini-whiteboard. Ask students to describe and explain the particle changes that occur in some of the chemical and physical changes. Share answers as a group, addressing any misconceptions and highlighting some key observations to look out for and any common errors.

Main activity

Activity: Comparing chemical and physical changes (20–25 mins) Show students the question:

Compare the particle changes in rusting with those that occur when iron melts. **(6 marks)**

As a class, discuss how to complete the 'Knowledge of' planning grid. Discuss the students' 'Knowledge of task' (the students' understanding of what the task is asking), 'Knowledge of strategies' (the different ways they can go about completing the task), and their 'Knowledge of self' (how confident they are with the subject knowledge the task requires). Model how to use the planning grid and some example phrases students may want to use. Students should then complete their 'Knowledge of' planning grid and use this to help to answer the exam question.

Review and reflect

Link back to the Trigger interest section and ask students to *think pair share* the question: Compare dissolving sodium carbonate in water and stirring baking powder into water (like in cake batter). Ask students to explain how we know if physical or chemical changes are occurring in each case. Discussions could be had about observations that could be made or activities that could be carried out to support explanations. For example: Boil off the water again to form just sodium carbonate – a reversible physical change. Show carbon dioxide gas (a new substance) has been formed – turns limewater cloudy – an irreversible chemical change. Compare the particle changes in each case.

Homework

Students write a short review of the lesson, including detailing what they learnt and the different strategies they encountered, including ones they chose not to use, as well as the struggles they overcame and those they still need to work on.

3.3 Apply: Reactants and products

Themes: Chemical reactions, Organic chemistry, Using resources

Sub-themes: Chemical equations, Fuels, Hydrocarbons, Reactions of organic compounds, Understanding chemical reactions

Chemistry NC links

- representing chemical reactions using formulae and using equations
- combustion, thermal decomposition, oxidation and displacement reactions

Working scientifically NC links

- make predictions using scientific knowledge and understanding
- use appropriate techniques, apparatus, and materials during fieldwork and laboratory work, paying attention to health and safety
- interpret observations and data, including identifying patterns and using observations, measurements and data to draw conclusions

Learning objectives

- Predict the products of combustion reactions.
- Interpret and write word equations.

Year 7 and 8 learning objectives to be applied

- Predict the products of the combustion of fuels.

Application

- Planning an investigation into identifying the products of combustion

Tier 2 vocabulary	Tier 3 vocabulary
carbon neutral, combustion, product	biofuel, hydrocarbon, reactant

SB in-text question answers

A Carbon dioxide and water **B** Octane and oxygen **C** Reactants – nitrogen and oxygen; product – nitrogen dioxide

Lesson plan

Reactivate knowledge

1 What is combustion?
2 What does the arrow mean in a word equation?
3 Write the chemical formula of nitrogen monoxide.

Lesson resources

Practical: *Identifying the products of combustion* (information sheet, student sheet, support sheet, teacher and technician sheet)

Answers: 1 A chemical reaction in which a substance reacts with oxygen, and energy is transferred to the surroundings as light and heat **2** Reacts to make **3** NO

Trigger interest

Ask students if they have ever seen a fire dancer, perhaps at a festival or on television. Ask them to *think pair share* the name of the substance that is burning and the names of any other fuels they are aware of.

Exposition of main content

Recalling the products of combustion Describe fuels as containing hydrocarbons and elicit the meaning of the term as substances that contain *only* hydrogen and carbon atoms. Ask students to recall the products of combustion of hydrocarbons and the tests for these substances (water – cobalt chloride paper turns pink; carbon dioxide – limewater turns cloudy).

Writing word equations As a class, discuss what word equations illustrate. Elicit the meanings of the terms *reactants* and *products*, and what the arrow represents. On mini-whiteboards, students practise writing word equations using the questions shown on the board. *Think pair share* strategies for writing word equations:

- key word and phrase use (e.g., *reacts to make/ produces/is formed by*)
- underlining/labelling reactants and products
- identifying the type of reaction that is occurring (e.g., combustion/oxidation/neutralisation) and any associated general reactions (e.g., acid + base ⟶ salt + water)
- filling in any 'missing' reactants/products (e.g., oxygen in combustion reactions)
- checking elements in reactants and products are consistent

Address any misconceptions.

Main activity

Practical: Identifying the products of combustion (25–35 mins)

Writing risk assessments Students examine the information given to them about the method and substances being used in the class practical and *think pair share* ideas for a risk assessment. Discuss what the hazards are, the risks, and the control measures. Students then individually write risk assessments for the practical.

Identifying the products of combustion
Students burn a candle under a gas jar and then test the products to determine if water and carbon dioxide are present. They record their observations in an appropriate table and then answer the questions that follow.

Review and reflect

Students are shown examples where the description of a reaction has been turned into a word equation that has some errors (e.g., reactants and products the wrong way round, missing substances, a equals sign) and they have to spot and correct the errors, explaining why they were wrong.

Homework

Students research an alternative fuel, and then write a short paragraph comparing their alternative fuel to paraffin (candle wax). They should compare the sources of these fuels, their availability for use, and the product(s) of combustion of these fuels.

3.4 Apply: Writing chemical equations

Themes: Bonding and properties of materials, Chemical reactions

Sub-themes: Chemical equations, Chemical formulae

Chemistry NC links

- chemical symbols and formulae for elements and compounds
- chemical reactions as the rearrangement of atoms
- representing chemical reactions using formulae and using equations

Learning objective
• Interpret and write balanced symbol equations.

Year 7 and 8 learning objectives to be applied	Application
• Describe what a balanced formula equation shows.	• Using metacognitive strategies to write balanced symbol equations

Tier 2 vocabulary	Tier 3 vocabulary
balanced	symbol equation

SB in-text question answers
A Reactants – lithium and oxygen; product – lithium oxide **B** 4 **C** CO_2 and H_2O

Lesson plan

Reactivate knowledge

1 What are products in chemistry?

2 What is a ratio?

3 What is the chemical formula of water?

Lesson resources

Activity: *Writing balanced symbol equations* (student sheet, teacher and technician sheet)

Answers: 1 The substances that are made in a chemical reaction **2** A way of comparing values **3** H_2O

Trigger interest

Show students the image of electric scooters in Figure 1 of the Student Book and inform them that on New Year's Day 2022, fire damaged 200 electric scooters and their batteries in a Bristol warehouse. Ask students to suggest why the batteries burnt. Explain that scooter batteries contain the element lithium. Ask students to recall information about lithium and its reactivity (it is in Group 1, it is very reactive, it reacts rapidly with water to produce lithium hydroxide). Explain that lithium catches fire easily, so a small spark was enough to start the blaze.

Exposition of main content

Writing equations As a class, discuss how to write word equations and ask students to suggest the word equation for the combustion of lithium to produce lithium oxide (lithium + oxygen ⟶ lithium oxide). Explain that word equations can be useful, but balanced symbol equations can give us more information. Show the balanced symbol equation for the combustion of lithium and ask students to suggest what extra information it offers. Elicit that it shows the formulae of the reactants and products in the reaction and the ratio of their amounts.

Chemical formulae Ask students to recall what a chemical formula shows (the ratios of atoms of each element in a substance, with the small subscript number showing the ratios, and no number indicating a ratio of 1), and to interpret the chemical formulae on the board. Impress on students that the chemical formula of a substance cannot be changed: it shows what a substance is made of and the ratios of elements in the substance.

Main activity

Introducing balancing techniques Introduce some different strategies the students may choose to use to write balanced symbol equations, with or without mini-whiteboards: the 'ping-pong' method, seesaw method, physical modelling using beads/counters, and 'grouping and protecting' method. For each method, model how to use it and then attempt a short question as a class. Students may want to copy an example of some of the methods into their books.

Activity: Writing balanced symbol equations (15–20 mins)

Show students the question:

Write the balanced symbol equation for the combustion of pentane, C_5H_{12}*, to form carbon dioxide and water.* **(3 marks)**

Introduce the *comprehension, connection, strategies, and reflection* metacognition activity, and briefly outline how to complete it on the activity sheet.

As a class, discuss students' comprehension of the task (their understanding of what the task requires), their connection to it (similar tasks they have completed before), and strategies they could use to complete it (different methods for completing the task and their confidence in using them). Students should then complete the comprehension, connection, and strategy sections of the activity sheet, before using their chosen method to answer the question.

Students should then use their comprehension, connection, strategies, and reflection grid to complete the questions that follow.

Review and reflect

Students should complete the reflection section of their activity sheet, considering if their comprehension of the task was correct, whether the connections they made were accurate, and whether the strategies they used were suitable.

Homework

Students should create a 'How to', for a Year 7 or 8 student explaining how to balance equations. This could be in the form of a leaflet, poster, or a simple bullet pointed list of instructions. They should include labelled diagrams and examples where possible.

3.5 Apply: Conservation of mass

Themes: Bonding and properties of materials, Chemical analysis, Chemical reactions

Sub-themes: Chemical equations, Chemical formulae, Quantitative analysis

Chemistry NC links

- chemical symbols and formulae for elements and compounds (KS3)
- conservation of mass, changes of state and chemical reactions (KS3)
- representing chemical reactions using formulae and using equations (KS3)
- balanced chemical equations, ionic equations and state symbols (KS4)
- quantitative interpretation of balanced equations (KS4)

Working scientifically NC links

- interpret observations and data, including identifying patterns and using observations, measurements and data to draw conclusions
- present reasoned explanations, including explaining data in relation to predictions and hypotheses

Learning objectives

- Apply the idea of conservation of mass.
- Apply ratios to calculate reacting masses.

Year 7 and 8 learning objectives to be applied

- Explain why the total mass of reactants is equal to the total mass of products in a chemical reaction.

Application

- Apply ratios to calculate reacting masses.

Tier 2 vocabulary	Tier 3 vocabulary
ratio	conservation of mass

SB in-text question answers

A 1 + 3 = 4 in reactants, 2 + 2 = 4 in products
B 1.00 + 0.40 = 1.40 g **C** 0.72 g × 100 = 72 g

Lesson plan

Reactivate knowledge

1 What does a balanced symbol equation show?
2 What is the chemical symbol for carbon?
3 Which elements do H and O represent?

Answers: 1 The reactants and products and their formulae, the ratio of the amounts of the reactants and products **2** C **3** H = hydrogen; O = oxygen

Lesson resources

Activity: *Calculating reacting masses* (student sheet, teacher and technician sheet)

Trigger interest

Show students the image of paracetamol in Figure 1 of the Student Book and ask what it is used for. Ask students how scientists know how to make paracetamol; elicit that they need to know how much of each reactant is needed to make the end product, that is, a 'recipe' for paracetamol.

Exposition of main content

Conservation of mass Show students the word equation, then the balanced symbol equation, for a reaction to produce paracetamol and count the atoms on each side to show it is balanced.

Demonstrations of conservation of mass Perform two demonstrations, the production of magnesium carbonate precipitate, and the reaction between hydrochloric acid and marble chips releasing carbon dioxide in the gas state. Use these demonstrations to show the law of conservation of mass – the total mass of reactants is equal to the total mass of products. The hydrochloric acid and marble chips demonstration is performed in a sealed bag to stop any gas from escaping (this would cause the mass of the products to appear less than the reactants). Ask students about what would happen if the bag was unsealed.

Main activity

Activity: Calculating reacting masses (25 mins) Provide mini-whiteboards so students can make notes/make mistakes and write down calculations they may need during the activity. Alongside class discussion, students should complete Tasks 1–3 on the activity sheet. As a class, discuss the use of ratios in baking recipes, and how students are usually comfortable with doubling/halving the masses of the ingredients (reactants) in recipes to make more or less of a pudding/cake (product). Discuss how the ratio of the masses needs to stay the same, otherwise we wouldn't be making the same cake. Discuss how this relates to making substances in science – the use of ratios of masses of our reactants (ingredients) and products (pastry) helps us to scale up or down reactions and makes sure we produce the mass of products we want. Apply this to the example of producing paracetamol.

Students then practise using ratios and calculating reacting masses using the questions on the activity sheet.

Review and reflect

Students should complete a Plus Minus Interesting (PMI) grid in their books detailing something that went well in the lesson, something they struggled with or found more complex, and something they found interesting, to reflect their learning in this lesson.

Homework

Students should write a step-by-step method for calculating reacting masses using an example from the lesson.

3.6 Apply: Energy in chemical reactions

Themes: Atomic structure and the particle model, Chemical reactions, Energy changes and rates of reactions

Sub-themes: Energy changes in chemical reactions, States of matter, Understanding chemical reactions

Chemistry NC links

- combustion, thermal decomposition, oxidation and displacement reactions
- exothermic and endothermic chemical reactions (qualitative)

Working scientifically NC links

- apply mathematical concepts and calculate results
- present reasoned explanations, including explaining data in relation to predictions and hypotheses
- evaluate data, showing awareness of potential sources of random and systematic error
- use and derive simple equations and carry out appropriate calculations

Learning objectives

- Compare energy changes in combustion reactions.
- Analyse results and make conclusions.

Year 7 and 8 learning objectives to be applied

- Identify exothermic and endothermic changes from temperature data.

Application

- Calculating the energy content in food by combustion.

Tier 2 vocabulary	Tier 3 vocabulary
combustion, energy	glucose

SB in-text question answers

A To **B** Carbon dioxide and water **C** So that the test is fair

Lesson plan

Reactivate knowledge

1 What is combustion?

2 What is the unit of energy?

3 What is the unit of temperature?

Lesson resources

Practical: *Energy content in food* (student sheet, support sheet, teacher and technician sheet)

Answers: 1 A chemical reaction in which a substance reacts with oxygen, and energy is transferred to the surroundings as light and heat **2** Joule, J **3** Degrees Celsius, °C

Trigger interest

Show students the image of burning sweets in Figure 1 of the Student Book and ask them to suggest what is happening. Ask students to suggest how we know a reaction is occurring and to *think pair share* what transfers of energy are occurring during the experiment. Explain that the chemical energy contained in the sweets is transferred to the surroundings as heat energy when it combines with oxygen from the air.

Exposition of main content

Why do we need food? Ask students to suggest why we eat food and why athletes 'carb-load' before big events and races. Elicit that food is a source of energy, and that glucose, a type of sugar found in sweets and other foods, is needed to enter cells and combine with oxygen to release energy to allow the cells to function. Explain that the reaction between glucose and oxygen in cells (known as respiration) is similar to a combustion reaction, and ask students to suggest the products (carbon dioxide and water) – students may already know this is the equation for respiration (glucose + oxygen ⟶ carbon dioxide + water). Link back to athletes 'carb-loading' and discuss how different foods can transfer different amounts of energy.

Energy transfer Students *think pair share* how we could test how much energy is transferred by different foodstuffs and then briefly describe the practical. Introduce the calculations students are going to use to calculate the amount of energy transferred from the foodstuff to the water through heating and to calculate the amount of energy per gram of food.

Main activity

Practical: Energy content in food (25 mins)
Allocate foodstuffs to students. Students then write a hypothesis that can be tested in the practical. They then perform the practical using burning foodstuffs to heat 10 cm^3 of water, calculating the approximate amount of energy transferred during heating and the energy per gram of burning food, and recording their results in a suitable table. Students then write a conclusion for the experiment summarising their results and commenting on whether their results support or disprove their hypothesis.

Review and reflect

Share students' results as a class and discuss any discrepancies between students who tested the same foodstuffs. Then, in groups, students should discuss why they think different results were obtained, the parts of the method that were designed to keep the results as accurate as possible (using 10 cm^3 of water each time, refreshing the test tube so it doesn't pre-emptively heat up the new water, holding the food 1 cm from the bottom of the test tube, recording the mass of food burnt each time), and where some sources of error lie (measuring the mass of food burnt, holding the burning food too close/far away, losing heat to the environment).

Homework

Students should write a paragraph discussing their findings and whether or not they support their hypothesis using the data collected in their experiment. They should comment on how reliable the results are and any sources of error, and should include ideas for improving the practical method for next time.

3.7 Reach: Exothermic and endothermic

Themes: Chemical reactions, Energy changes and rates of reactions

Sub-themes: Energy changes in chemical reactions, Understanding chemical reactions

Chemistry NC links

- combustion, thermal decomposition, oxidation and displacement reactions (KS3)
- exothermic and endothermic chemical reactions (qualitative) (KS3)
- changes of state of matter in terms of particle kinetics, energy transfers and the relative strength of chemical bonds and intermolecular forces (KS4)
- measurement of energy changes in chemical reactions (qualitative) (KS4)

Working scientifically NC links

- pay attention to objectivity and concern for accuracy, precision, repeatability and reproducibility
- select, plan and carry out the most appropriate types of scientific enquiries to test predictions, including identifying independent, dependent and control variables, where appropriate
- make and record observations and measurements using a range of methods for different investigations; and evaluate the reliability of methods and suggest possible improvements

Learning objective

- Interpret observations and data to decide if a change is exothermic or endothermic.

Year 7 and 8 learning objectives to be built on

- Describe the energy changes in exothermic and endothermic changes.

Key Stage 4 link

- Exothermic and endothermic reactions

Tier 2 vocabulary	Tier 3 vocabulary
neutralisation, observation, solution	endothermic, exothermic

SB in-text question answers

A A change in which energy is transferred from the surroundings **B** 63 – 21 = 42 °C **C** Potassium nitrate

Lesson plan

Reactivate knowledge

1 What is an exothermic change?
2 What is a neutralisation reaction?
3 What is a solute?

Answers: 1 A change in which energy is transferred to the surroundings **2** A reaction in which an acid is 'cancelled out' by a base **3** A substance that dissolves to make a solution

Lesson resources

Practical: *Exothermic or endothermic?* (student sheet, support sheet, teacher and technician sheet)

Trigger interest

Show the image of fireworks in Figure 1 of the Student Book and ask students to describe the reaction in terms of energy changes. The chemical reactions in fireworks transfer energy to the surroundings in the form of heat, light, and sound. Ask students to name the type of reaction that occurs when fireworks go off (combustion). Elicit that combustion is an exothermic reaction.

Exposition of main content

Transfers of energy Ask students to recall the differences between exothermic and endothermic changes, and, if possible, to give examples of each. Elicit that endothermic changes transfer energy *from* the surroundings to the substances that are reacting, changing state, or dissolving. Probe students to recall which changes of state are endothermic (melting and evaporating). Elicit that exothermic changes transfer energy from the substances that are reacting, changing state, or dissolving, *to* the surroundings. Probe students to recall which changes of state are exothermic (freezing and condensing).

Students should *think pair share* ways they can determine if a change is endothermic or exothermic. Ask what observations and measurements they could take in a practical to establish this (measure temperature change with hands/a thermometer).

Main activity

Writing a plan As a class, write a checklist of requirements for a good, clear, well-written practical plan. Students should suggest key content that needs to be included, and any examples that will support other learners (this could include good or bad examples). This should be written on the board to be used in the practical. Discussions may be needed about what the independent variable (the substances reacting), dependent variable (temperature), and control variables (volumes of solutions, equipment used) are.

Practical: Exothermic or endothermic? (20–25 mins) In small groups, students follow the checklist to design and write a short practical plan for determining if two processes (the neutralisation reactions between sodium hydroxide and hydrochloric acid and between sodium hydrogen carbonate and citric acid) are endothermic or exothermic changes. They should include an equipment list, a method, and an appropriate results table. The risk assessment is provided. Groups then swap plans and make suggestions for improving the method and plan, before swapping back and implementing those changes. Finally, students follow a third group's plan, determining whether the processes are exothermic or endothermic, and analyse the plan they were given. As a class, discuss results, what worked and what didn't about the different plans, and how to add to and improve on the content checklist.

Review and reflect

Students write exothermic and endothermic on opposite sides of a mini-whiteboard, then read through the information on the board and determine if the change described is endothermic or exothermic. Use targeted questioning to query students' answers to check their explanations and address any misconceptions.

Homework

Students should write a paragraph to describe how they minimised sources of error in the practical procedure they performed in the lesson, and suggest and explain improvements to the practical procedure to reduce errors further.

C3 Chapter 3: Chemical changes

Introduction to checkpoint intervention

This checkpoint intervention provides suggestions for a lesson to follow up the C3 3 *Chemical changes* checkpoint assessment. Depending on the outcomes of the assessment, these suggestions could move a class towards secure, to target specific learning outcomes, or to consolidate knowledge for students achieving secure outcomes. Use students' outcomes from the checkpoint assessment to address general misconceptions from the content covered. Be prepared to re-cover content as required; three key concepts from the chapter are covered in more detail below.

Checkpoint secure learning outcomes

- Compare particle changes in physical and chemical changes.
- Predict the products of combustion reactions.
- Interpret and write balanced symbol equations.
- Apply the idea of conservation of mass.
- Compare energy changes in combustion reactions.
- Interpret observations and data to decide if a change is exothermic or endothermic.

Student reflection

Encourage students to reflect on whether there were any checkpoint questions they found difficult or straightforward, as well as their preparation for the checkpoint assessment, including how they revised, the time spent on revision, and what areas they could improve on. For more strategies, see the Metacognition chapter or the Metacognition in Key Stage 3 Science guide on Kerboodle.

Intervention activity: Interpret and write balanced symbol equations

This can be very difficult to learn to begin with and many students can switch off with it seeming impossible. Choosing simple examples to begin with is the key to building a feeling of success. We would recommend starting with equations where at least one reactant or product is an element and building up from there. While students are gaining confidence, try to avoid examples with polyatomic ions, for example, SO_4. Experienced chemists treat the sulfate as a single unit, which lowers demand on working memory. Students see it as one sulfur and four oxygens and this can lead to cognitive overload.

Some simple examples to start with include:

- $Na + Cl_2 \rightarrow NaCl$
- $Mg + O_2 \rightarrow MgO$ this is very similar to the previous example, so students can try independently after their teacher has shown them the sodium chloride example.
- $H_2 + Br_2 \rightarrow HBr$
- $Al + Br_2 \rightarrow AlBr_3$
- $S + O_2 \rightarrow SO_3$
- $H_2 + O_2 \rightarrow H_2O$
- $Zn + HCl \rightarrow H_2 + ZnCl_2$

Intervention activity: Apply the idea of conservation of mass

The conservation of mass is extremely obvious for macroscopic systems, for example, making a sandwich from ingredients. However, with chemical systems, particularly those involving gases, it can cause much confusion. It is best to introduce this topic with precipitation reactions (e.g., lead nitrate and potassium iodide, starch and iodine, copper sulfate and sodium hydroxide) before introducing examples involving gases. It is important to emphasise that the conservation of mass is still true for reactions involving gases, it is just that gases usually escape the reaction system. Plenty of examples of these are provided in the pinchpoint activity.

Intervention activity: Interpret observations and data to decide if a change is exothermic or endothermic

Although many students do not seem to struggle with memorising '*exothermic* = *gets hotter*' and '*endothermic* = *gets colder*', relating this to energy transfer between the system and surroundings often creates a stumbling block. One potential way around this is to focus on the word exothermic and highlighting to students that -*thermic* is *thermal energy* and not temperature. Therefore, students will be able interpret *exothermic* as "*exiting thermal energy*". Since the thermal energy is exiting the system (i.e., the chemicals themselves), the surrounding (e.g., the test tube or their hand) feel warmer. The prefix *endo*- means *within*, which is less obvious, so we recommend starting with exothermic.

Review how energy transfers are taught in P2 2 *Energy* and this will help connect with what students already know about energy.

For example (the numbers are completely made up):

Chemical equation	ethanol + oxygen	→	carbon dioxide + water	
Total energy in J	150	→	150	This stays the same since energy is always conserved.
Amount of thermal energy in J	50	→	110	Thermal energy has increased in the products. Therefore the surroundings are at a higher temperature, so temperature rises.
Amount of chemical energy in J	100	→	40	This is total energy – thermal energy. You can clearly see the amount of chemical energy has gone down (by the same amount as the thermal energy has increased). Energy has exited the system, to the surroundings.

C3 Chapter 4: Useful chemical reactions

Introduction to chapter

This chapter explores further chemical reactions, with a focus on metals and the reactivity series. Students will first cover metals and their properties. They then study the reactivity series, displacement reactions, and catalysts before concluding the chapter with some more mathematical topics, including relative mass and yields.
This chapter provides excellent opportunity for practical work and practising maths skills in science.

Core concepts

- Materials
- Word equations

Prerequisite knowledge

- Chemical reactions as the rearrangement of atoms
- Representing chemical reactions using formulae and using equations
- Combustion, thermal decomposition, oxidation and displacement reactions
- Reactions of acids with metals to produce a salt plus hydrogen
- What catalysts do
- The properties of metals and non-metals
- The order of metals and carbon in the reactivity series
- The use of carbon in obtaining metals from metal oxides.

GCSE links

- characteristic properties of metals and non-metals
- chemical reactivity of elements in relation to their position in the Periodic Table
- the chemistry of acids; reactions with some metals and carbonates
- electrolysis of molten ionic liquids and aqueous ionic solutions
- extraction and purification of metals related to the position of carbon in the reactivity series.

Key learning misconceptions

- The reactivity series only applies to metals.
- All metals are the same
- Chemical reactions are events
- Metals are dug out the ground
- Catalysts don't take part in chemical reactions

Teaching preparation

Research skills

Using metals This provides an excellent opportunity for students to research different metallic element of the Periodic Table. However, given that students are novices, they will find it very difficult to research information accurately and present it in a meaningful way. Without guidance, you might be left with copied and pasted information from the internet of which the student has little or no understanding. It would be better to provide students with an element to research (you can pick based on how much you wish to stretch or support the student) and some initial questions for them to find the answers to. For example, *When was it discovered? What is its chemical symbol? Why does it have the name that it does?* This can then lead into more open-ended questions, for example, *Does your element have any uses in medicine? Does it have any uses in your home?*

Another opportunity for independent research is the catalyst topic.

Key maths skill

Relative mass calculations The end of this chapter is highly mathematical. The best ways for students to improve at this are clear explanations, teacher modelling, and plenty of practice. Although these ideas will be developed at GCSE level, it is not quite the same way in which reacting mass calculations are dealt with at GCSE (e.g., there are no moles). If you are new to teaching this topic, it is worth completing the activity questions yourself, complete with working, to ensure that you can foresee any pitfalls that the students might encounter.

Teaching strategy

Atomic scales The reactivity series underpins a large chunk of this chapter and will be covered again at GCSE. The sooner students memorise the reactivity series the better. A good approach is to be upfront with them about this and tell them you will test them on it every lesson for the next week. This makes expectations crystal-clear to students and provides them ample opportunity to practise recalling that information. As for memory strategies, students should be encouraged to come up with a mnemonic to help them remember the order. Note: students should also be able to place carbon on the series even though it is not a metal.

Curriculum coherence

Maths: Mathematics is an integral part of science at GCSE and beyond (it is 20% of the marks in GCSE exams for chemistry). It is worth arranging a meeting between the science and mathematics teachers to ensure that across the school there is parity of explanations and understanding of numerical pedagogy.

Key maths skill

Percentages Although fundamentally straightforward, percentages can be challenging for students. It is worth stressing that since the actual yield will always be less than the theoretical yield, the percentage yields will always be less than 100%. This can help students catch mistakes they make when calculating the percentage yield.

4.1 Reactivate: Useful chemical reactions

Theme: Chemical reactions

Sub-theme: Reactivity series

Chemistry NC links

- chemical reactions as the rearrangement of atoms
- representing chemical reactions using formulae and using equations
- combustion, thermal decomposition, oxidation and displacement reactions
- what catalysts do

Working scientifically NC links

- understand that scientific methods and theories develop as earlier explanations are modified to take account of new evidence and ideas, together with the importance of publishing results and peer review
- use appropriate techniques, apparatus, and materials during fieldwork and laboratory work, paying attention to health and safety

Learning objective

- Predict whether or not pairs of substances react in displacement reactions.

Year 7 learning objectives to be revised

Year 8 learning objectives to be revised

- Predict whether a pair of substances would or would not react in a displacement reaction.

Pinchpoint focus

- Reactivity series and displacement reactions.

Tier 2 vocabulary	Tier 3 vocabulary
chemical, displace, displacement, physical, property, reactive	displacement reaction

Trigger interest

Ask: *Why are pipes made out of copper and not sodium?*

Ask students to recall the reactivity series and a funny sentence they use to help remember the series.

Recall

Ask students to recall the reactivity series. Ensure they know the series well.

Provide students with some reactions and ask them to determine if the reaction happens and if so, what the products are.

Main activity

Pinchpoint: Useful chemical reactions (20–25 mins)

The reactivity series allows students to predict if certain chemical reactions will happen. This will lead students to be able to determine how certain metals are extracted.

Students answer the pinchpoint question and complete the follow-up activity with the same letter. They then complete the *What I need to remember* activity. Information about each incorrect answer's misconceptions can be found on the teacher sheet. You may want to divide the class into groups, depending on the letter answer chosen, to support each other when completing the follow-up activity.

Students carry out a series of displacement reactions on spotting tiles.

Review and reflect

Students are shown the pinchpoint question again and asked to reflect on whether they chose the correct letter and if they would change their answer now. If they think they got it wrong, ask them to think about why they chose that answer.

Reveal the correct answer. Review with students why it is correct and the other options are not. Guidance can be found on the teacher sheet. Students could be asked to pick an incorrect answer and write an explanation to a fictional student who chose it, explaining why it cannot be correct.

Ask students how verifiable the results from a displacement experiment are. Encourage students to ask if, for example, magnesium will displace copper from copper chloride, copper sulfate, and copper nitrate.

Homework

Students fill in the incomplete knowledge organiser.

4.2 Apply: Using metals

Themes: Bonding and properties of materials, The Periodic Table

Sub-themes: Metals and alloys, Metals and non-metals

Chemistry NC links

- representing chemical reactions using formulae and using equations
- the varying physical and chemical properties of different elements
- the properties of metals and non-metals
- the order of metals and carbon in the reactivity series

Working scientifically NC links

- make predictions using scientific knowledge and understanding
- present reasoned explanations, including explaining data in relation to predictions and hypotheses

Learning objective

- Explain how metal properties make them suitable for their uses.

Year 7 and 8 learning objective to be applied

- Use a pattern to predict the products of the reaction of a metal with an acid

Application

- Research and present information on unfamiliar metals

Tier 2 vocabulary	Tier 3 vocabulary
hardness, malleable, reactive, unreactive	ductile

SB in-text question answers

A chromium **B** chromium oxide **C** For example, freshly cut chromium reacts with oxygen from the air, but palladium does not react with oxygen

Lesson plan

Reactivate knowledge

1 What are physical properties?
2 What are chemical properties?
3 What does 'reactive' mean in chemistry?

Lesson resources

Activity: *Uses of metals* (information sheet, student sheet, teacher and technician sheet)

Answers: 1 Properties that you can observe or measure without changing a material **2** Properties that describe the chemical reactions of a substance **3** A substance is reactive if it easily takes part in chemical reactions

Trigger interest

Show students the three images on the Lesson presentation of objects made from or coated in metal. Ask: *Why have these metals been chosen?* Discuss how and why the objects are used, along with what conditions and stresses they need to be able to withstand. This will be returned to at the end of the lesson.

Exposition of main content

Introduction to different metals As a class, recall the properties of 'typical' Group 1 and 2 metals (soft, malleable, ductile, shiny, good thermal and electrical conductors, low density, sonorous). Ask if these properties are consistent for all metals (no, e.g., most everyday metals are hard, not soft). Discuss how all metals have some of the same properties (malleable, ductile, lustrous, thermal and electrical conductor) but that there is still some variety (e.g., melting point). Ask students to suggest any metals whose properties differ from the 'typical' metals (e.g., mercury – in the liquid state at room temperature).

How do we choose materials? Ask students to recall the meanings of the terms *chemical properties* and *physical properties*. As a class, recall and discuss why it is important we use the right materials for different uses. Elicit that the right material for a use will mean there is increased efficiency and safety, as well as decreased cost whilst allowing the object to look and perform at its best.

Word equations Very briefly recap how to write and interpret word equations, including how to identify and define reactants and products, and what the arrow means.

Main activity

Cornell note-taking and methods for extracting information Ask students to *think pair share* any methods for note-taking that they are aware of or have used before. Discuss the pros and cons of each method as a class, giving examples or modelling each method suggested if needed.

Activity: Uses of metals (15–20 mins) In groups of five, assign each group member a different metal (aluminium, copper, gold, silver, platinum). Students should read through the information sheet for their metal and extract information about its properties and its uses. Students make notes and summarise the information about their assigned metal in their books using whichever method they choose (e.g., create a table to fill in, bullet point list, flow maps). Students then present their summarised information back to their group members, each taking it in turn to pass on the information succinctly and concisely.

Individually, students then answer questions about the metals, explaining how the physical and chemical properties of each metal make it suitable for its use(s) and writing word equations for some reactions of these elements.

Review and reflect

Link back to the Trigger interest section and show the images of the metals making surgical tools and coating objects on the Lesson presentation. Use targeted questioning to develop and check understanding and to ensure students can explain why these metals are used here and why their properties make them suitable for their uses. (Chromium is hard and shiny, so is used to coat objects. Palladium is used in surgical instruments since it is ductile, strong, and unreactive with a high melting point. More information can be found in the Student Book.)

Homework

Students should research the chemical and physical properties of a metal element they are not as familiar with (some good examples could be: tin, lead, mercury, ruthenium, tungsten, zirconium, cadmium). They should choose a note-taking method that they are not as familiar with and use it write a short paragraph explaining why the properties of that element make it suitable for its use. Remind students this is practice at independent working: they should not be copying websites word-for-word, but making their own notes.

4.3 Apply: The reactivity series

Theme: Chemical reactions

Sub-themes: Reactivity series

Chemistry NC links

- chemical reactions as the rearrangement of atoms
- combustion, thermal decomposition, oxidation and displacement reactions
- reactions of acids with metals to produce a salt plus hydrogen

Working scientifically NC links

- evaluate risks
- make predictions using scientific knowledge and understanding
- select, plan and carry out the most appropriate types of scientific enquiries to test predictions, including identifying independent, dependent and control variables, where appropriate
- use appropriate techniques, apparatus, and materials during fieldwork and laboratory work, paying attention to health and safety
- make and record observations and measurements using a range of methods for different investigations; and evaluate the reliability of methods and suggest possible improvements

Learning objective

- Determine the position of a metal in the reactivity series.

Year 7 and 8 learning objective to be applied

- Use the reactivity series to predict reactions

Application

- Investigating the reactions of metals with hydrochloric acid

Tier 2 vocabulary	Tier 3 vocabulary
hazard	

SB in-text question answers

A Gold **B** Left – a health hazard that causes damage over time; right – dangerous to the environment **C** Iron

Lesson plan

Reactivate knowledge

1 What is the reactivity series?

2 What is a hypothesis?

3 What are control variables?

Answers: **1** A list of metals in order of how vigorously they react with other substances **2** A possible explanation that can be tested further **3** Variables to keep constant in an investigation

Lesson resources

Practical: *Comparing reactivity of metals* (student sheet, teacher and technician sheet)

Trigger interest

Show students the images in Figure 1 of the Student Book, or conduct a demonstration, of metals (magnesium and calcium) reacting with water. Ask: *Which metal is higher in the reactivity series? How do you know?* (Calcium, as it is producing more bubbles per second, or bubbling more vigorously, than magnesium.)

Exposition of main content

The reaction of metals and acids Ask students to recall the products of the reaction between metals and acids. Elicit that hydrogen is produced in the reaction and ask students to recall the test for hydrogen (hydrogen combusts with a squeaky pop when a lit splint is placed near). Write the general equation on the board (metal + acid → metal salt + hydrogen).

Recalling the reactivity series Ask students to recall what the reactivity series is and any mnemonics for remembering the order of metals in the reactivity series. As a class, discuss why the reactivity series is an important and useful tool for scientists and the general public to use.

Writing practical plans As a class, recall and discuss the basic components of a good practical plan (hypothesis to test, risk assessment, identification of variables, method, results, and conclusion), giving short examples or clarifications if necessary.

Main activity

Practical: Comparing reactivity of metals (20–25 mins) Students plan, write, and conduct a short practical to determine the order of reactivity of five metals with dilute hydrochloric acid. Information is provided for students to use to write their risk assessments. Students then collect the substance in the gas state produced when the metals and acid react, and perform the squeaky pop test to prove this substance is hydrogen.

Students should then individually answer the questions that follow.

Review and reflect

In their practical groups, students discuss their results and the conclusions they drew from them, as well as any potential sources of error. Each group member then joins up with one group member from other groups and they discuss their results and conclusions – highlight any discrepancies and discuss how these may arise. Share these conclusions as a class.

Homework

Students should research and then create an A4-poster on the reactivity of metals. They should include the reactivity series, a mnemonic to remember it, information about the reactivity series and its use, and the reactivity and uses of metals.

4.4 Apply: Displacement reactions

Themes: Chemical reactions

Sub-themes: Reactivity series

Chemistry NC links

- chemical reactions as the rearrangement of atoms
- combustion, thermal decomposition, oxidation and displacement reactions

Working scientifically NC links

- evaluate risks
- make predictions using scientific knowledge and understanding
- select, plan and carry out the most appropriate types of scientific enquiries to test predictions, including identifying independent, dependent and control variables, where appropriate
- use appropriate techniques, apparatus, and materials during fieldwork and laboratory work, paying attention to health and safety
- make and record observations and measurements using a range of methods for different investigations; and evaluate the reliability of methods and suggest possible improvements

Learning objective

- Predict whether or not pairs of substances react in displacement reactions.

Year 7 and 8 learning objective to be applied

- Predict whether a pair of substances would or would not react in a displacement reaction

Application

- Interpret observations of displacement reactions in investigations.

Tier 2 vocabulary	Tier 3 vocabulary
	displacement reaction

SB in-text question answers

A Copper & zinc oxide and copper & iron oxide
B Zinc oxide and iron **C** Zinc sulfate

Lesson plan

Reactivate knowledge

1 Look at Figure 2. Which is more reactive, zinc or copper?
2 What does the arrow mean in a chemical equation?
3 What is a conclusion?

Answers: 1 Zinc **2** Reacts to make **3** A summary and explanation of observations or data that were collected to answer a scientific question

Lesson resources

Practical: *Identifying displacement reactions* (student sheet, teacher and technician sheet)

Trigger interest

Tell students that railway tracks are joined together by liquid iron, which then cools and freezes, joining the two sections of rail together. But iron melts at about 1500 °C. This is too hot to melt in any normal way. Ask: *How do you think they get iron so hot?*

Show the image in Figure 1 of the Student Book of the thermite reaction and explain that the railway worker is using a chemical reaction called the thermite reaction to make liquid iron. Show the word equation explaining that aluminium is taking the place of iron in the compound, and ask students to recall what type of reaction this is (displacement reaction).

Thermite demonstration Perform a demonstration of the thermite reaction, the displacement of iron from iron oxide using aluminium. This should be practised beforehand or a video of the reaction can be shown instead.

Exposition of main content

Writing word equations for displacement reactions Students use mini-whiteboards to practise moving metals around to show displacement reactions.

Mistaken reactions Ask students to suggest what they will be looking out for to determine if a displacement reaction has occurred. Elicit that a displacement reaction will show a new metal appearing or a colour change in the metal present. The presence of iron could be tested with a magnet.

Main activity

Practical: Identifying displacement reactions (20–25 mins) Students predict, using the reactivity series, whether or not displacement reactions will occur. They then perform 16 different displacement reactions on a spotting tile, recording their observations in an appropriate table.

Students write word equations for the reactions they saw occurring and justify each reaction, or lack of reaction, using the reactivity series.

Review and reflect

Showing the practical grid on the board, 'slap' or cover up a square/tile. Students should then determine if a reaction would occur between the corresponding metal and metal compound solution. If so, they could write the word equation for that reaction on their mini-whiteboards, and if not, they write 'no reaction' and a brief explanation as to why not.

Homework

Students write up the experiment that they have completed today. This could be a piece of extended writing, a presentation, or a poster. Students should include what they did, their results, their conclusions, and any health and safety aspects.

4.5 Apply: Extracting metals

Themes: Chemical reactions

Sub-themes: Reactions of metals

Chemistry NC links

- chemical reactions as the rearrangement of atoms
- combustion, thermal decomposition, oxidation and displacement reactions
- the order of metals and carbon in the reactivity series
- the use of carbon in obtaining metals from metal oxides

Learning objective
• Explain why different methods are used to extract different metals from their ores.

Year 7 and 8 learning objective to be applied	Application
• Name the metals in the reactivity series that can be extracted by heating their oxides with carbon	• Extracting metals from their oxides using carbon

Tier 2 vocabulary	Tier 3 vocabulary
compare	electrolysis

SB in-text question answers
A Carbon dioxide **B** Magnesium or potassium **C** Chemistry/position in the reactivity series, cost

Lesson plan

Reactivate knowledge

Lesson resources
Practical: *Extracting copper and iron* (student sheets × 2, teacher and technician sheet)

1 What is an ore?
2 What is a displacement reaction?
3 Look at Figure 3. Which metals are more reactive than carbon?

Answers: 1 A rock that you can extract a metal from, and that has enough of the metal to make it worth extracting
2 A reaction in which a more reactive element/metal displaces, or pushes out, a less reactive element/metal from its compound
3 Potassium, sodium, magnesium, aluminium

Trigger interest

Ask students to share if they have ever had an X-ray and what they are used for. Discuss the hazards and health concerns surrounding getting X-rays. Using Figure 1 in the Student Book, ask students to suggest why X-ray technicians and radiographers need lead-lined aprons. Elicit that X-rays can damage living cells and that technicians and radiographers do hundreds a day. Therefore, they are much more exposed than those of us who go once to get an X-ray, so they need some extra protection.

Exposition of main content

Ores Ask students to recall what an ore is, where they are found, and what they are used for. Discuss how most reactive metal elements are found in compounds mixed with other compounds in rocks in the Earth's crust and those that have enough metal in to make it worth extracting are called ores.

If resources are available, give out samples of ores and the metal they make, (e.g., haematite and iron). Let students touch the samples and feel how an ore has different properties to the metal that we extract from it.

Extracting metals As a class, use a mnemonic (or two!) to recall the order of metals in the reactivity series, write it on the board and then add in carbon. Recall the definition of a displacement reaction (a reaction where a more reactive element/metal displaces, or pushes out, a less reactive element/metal from its compound). Recall and discuss that some metals are extracted from their ores using a displacement reaction and others using electrolysis. Briefly discuss the process of electrolysis (using electricity to separate substances).

Main activity

Practical: Extracting copper and iron (20–25 mins) Students extract copper from copper oxide and/or iron from iron oxide using carbon. They can demonstrate that copper has been produced due to its colour and they can show iron has been produced by holding a magnet nearby.

Students can complete one or both practicals depending on the availability of resources and timing.

Review and reflect

Showing the reactivity series on the board again, call out a metal. Students move to different sides of the room depending on whether that metal can be extracted from its ore using carbon or if electrolysis would be used. Students should be able to explain their answers. Mention very unreactive metals, such as gold or silver, and ask students to suggest methods of extraction for those too.

Homework

Students should create a knowledge organiser summarising their learning from the last few lessons, including identifying and explaining the methods for extraction of various metals in the reactivity series. Ask students to consider which method of organisation they are going to use and why they are going to use it.

4.6 Apply: Using metal: catalysts

Themes: Chemical reactions

Chemistry NC link

- what catalysts do

Sub-themes: Understanding chemical reactions

Working scientifically NC links

- use appropriate techniques, apparatus, and materials during fieldwork and laboratory work, paying attention to health and safety
- make and record observations and measurements using a range of methods for different investigations; and evaluate the reliability of methods and suggest possible improvements

Learning objective

- Explain how catalytic converters make exhaust products less harmful.

Year 7 and 8 learning objective to be applied

- Explain how the uses of elements are determined by their properties

Application

- Exploring the use of transition metals in catalytic converters

Tier 2 vocabulary	Tier 3 vocabulary
catalyst	catalytic converter, transition metal

SB in-text question answers

A Carbon monoxide, nitrogen monoxide, nitrogen dioxide **B** First reaction – carbon dioxide; second reaction – carbon dioxide and nitrogen **C** Palladium, platinum, and rhodium **D** Iron

Lesson plan

Reactivate knowledge

1 What is a hydrocarbon?
2 Name an important greenhouse gas.
3 In a chemical reaction, what are reactants, and what are products?

Lesson resources

Practical: *Colourful catalysts with cobalt* (student sheet, teacher and technician sheet)

Answers: 1 A compound whose molecules are made up of atoms of carbon and hydrogen only **2** Carbon dioxide/methane **3** Reactants – starting substances in a chemical reaction; products – substances that are made in a chemical reaction

Trigger interest

Show students the the catalytic converter in Figure 1 of the Student Book and ask if they know what it is and does. Explain that every night in March 2021, thieves in the UK stole 100 of these from under parked cars. Ask students to suggest why people might do this. Explain that catalytic converters contain valuable metals, which some people then steal to sell.

Exposition of main content

What is a catalyst? Explain that a catalyst speeds up a reaction, but is not used up in the reaction. Students may recall enzymes as catalysts in digestion. Most catalysts are made from transition elements, which are located in the middle of the Periodic Table.

Catalytic converters Show Figure 2 of the Student Book of the inside of a catalytic converter and inform students that it is made of a honeycomb structure of a ceramic material coated in valuable metals such as platinum, palladium, and rhodium. Ask students to recall what a ceramic material is and how it is made (a hard, brittle material made by firing a material, such as clay, at a high temperature). Ask students to find platinum, palladium, and rhodium on the Periodic Table and verify that they are transition elements.

Why do we need catalytic converters? Ask students to recall the products of combustion of fuel in cars (carbon dioxide and water), and any other substances they know are produced (e.g., nitrogen oxides). Discuss that carbon monoxide is also produced and elicit that it is a poisonous gas, as well as that oxides of nitrogen can cause respiratory problems and make rain acidic. Explain that catalytic converters change carbon monoxide and nitrogen oxides into gases naturally found in the atmosphere (carbon dioxide and nitrogen). As a class, write the word equations for these reactions on the board (carbon monoxide + oxygen → carbon dioxide; carbon monoxide + nitrogen monoxide → carbon dioxide + nitrogen). Let students vote on whether carbon dioxide, carbon monoxide, or nitrogen monoxide is least dangerous (carbon dioxide, although it's still a greenhouse gas). Ask a student to explain why they voted that way and allow students with opposing views can then argue their point. Ask students to re-vote and reflect on why they did or did not change their minds.

Main activity

Practical: Colourful catalysts with cobalt (10–20 mins) Students carry out a reaction of potassium sodium tartrate with hydrogen peroxide and a cobalt chloride catalyst. They record their observations and answer questions.

They should clearly see no visible reaction until the cobalt chloride is added. The cobalt chloride will turn from pink to green and back to pink again, showing that the cobalt(II) is not used up.

(Oxidation states of cobalt are beyond students at this level. Referring to it as pink cobalt and green cobalt might help reduce cognitive load.)

Review and reflect

Ask students to reflect on their learning over the last few lessons to identify what they think the main ideas are. If required, ask students to use sentence starters such as: "I feel knowledgeable about this topic because..."; "As a learner I have developed through this topic by..."; or "The critical knowledge for this topic is...".

Homework

Students should research another use of a catalyst, and write a paragraph describing its use, what the catalyst is made from, and how it is useful.

4.7 Reach: Relative mass

Themes: Bonding and properties of materials

Sub-theme: Chemical formulae

Chemistry NC link

- a simple model of the atom consisting of the nucleus and electrons, relative atomic mass, electronic charge and isotopes (KS4)

Working scientifically NC link

- recognising the importance of scientific quantities and understanding how they are determined (KS4)

Learning objective

- Calculate relative masses of molecular elements and compounds.

Year 7 and 8 learning objective to be built on

- Explain why the total mass of reactants is equal to the total mass of products in a chemical reaction

Key Stage 4 link

- Relative atomic and formula masses

Tier 2 vocabulary	Tier 3 vocabulary
	reacting mass, relative mass

SB in-text question answers

A $9 + 10 = 19$ **B** $14 + (2 \times 16) = 46$ **C** 44 g

Lesson plan

Reactivate knowledge

1 Name three sub-atomic particles.

2 What are the relative masses of protons and neutrons?

3 What does 'mass is conserved' mean?

Lesson resources

Activity: *Calculating relative mass* (student sheet, support sheet, teacher and technician sheet)

Answers: 1 Protons, neutrons, electrons **2** 1 **3** The total mass of products in a chemical reaction is equal to the total mass of reactants

Trigger interest

Ask: *Why is it important to know how much product you are producing?* Ask students to write their answers down on post-it notes and stick them to the board. Read out some of the answers.

Show the image of the boiler in Figure 1 of the Student Book and explain that some gas boilers burn methane to heat up water, which then flows into radiators to heat up our houses. Ask students to recall the word equation for the combustion of methane and to suggest how they could work out what mass of carbon dioxide their boiler is producing. Explain that this would be useful to compare the energy efficiencies and amounts of greenhouse gases produced by different boilers.

Exposition of main content

Relative mass Ask students to recall the three sub-atomic particles and their relative masses and locations. Recall that because the mass of an electron is so small compared to protons and neutrons, the majority of the mass of an atom is found in the nucleus; this is the mass we use to determine masses of atoms. As a class and using the table on the board, add the numbers of protons and neutrons in different elements to calculate the relative mass of an atom of each element, writing the equation and answers on the board.

Show chemical formulae of molecules on the board and ask students to suggest how we might determine the relative masses of these molecules. As a class, fill in the table to calculate the masses of these molecules.

Conservation of mass Students should be well-versed in the law of conservation of mass. Ask them to recall it and how it applies to balanced symbol equations. On mini-whiteboards, students should write the word equation and the balanced symbol equation for the combustion reaction of methane (methane + oxygen → carbon dioxide + water, $CH_4 + 2O_2 \rightarrow CO_2 + 2H_2O$). Elicit these from the students and write them on the board. Discuss the reacting ratios and ask students to calculate the masses of each molecule in the balanced symbol equation. Explain that if we show the masses of the substances in the equation (that is, taking into account the reacting ratios) we know the masses that will react together, called reacting masses. Ask students to calculate these and write them below the balanced symbol equation on the board. Ask students to suggest why this might be useful; lead them to the answer that we can therefore work out how much of each reactant we need to make a certain amount of product, or vice versa.

Main activity

Activity: Calculating relative mass (20–25 mins) Students should now be able to complete the task, filling in the blanks in the tables to determine, where necessary, relative mass of an atom of an element, the relative masses of each molecule, and the reacting masses in a balanced chemical equation.

Review and reflect

Students should calculate the relative masses of atoms and molecules shown on the board. If time allows, some students can then attempt to write their own. This may be completed on mini-whiteboards.

Homework

Students should complete a PMI (plus, minus, interesting) grid in their books detailing something that went well in the lesson, something they struggled with or found more complex, and something they found interesting, to reflect their learning in this lesson.

4.8 Reach: Calculating yield

Themes: Chemical analysis, Chemical reactions

Sub-themes: Chemical equations, Quantitative analysis

Chemistry NC links

- quantitative interpretation of balanced equations (KS4)
- balanced chemical equations, ionic equations and state symbols (KS4)

Learning objective

- Calculate percentage yields.

Year 7 and 8 learning objective to be built on

- Calculate the mass of one reactant or product in a chemical reaction, given the masses of all other reactants and products

Key Stage 4 link

- Calculate percentage yields.

Tier 2 vocabulary	Tier 3 vocabulary
	actual yield, percentage yield, theoretical yield

SB in-text question answers

A 202 g ÷ 100 = 2.02 g **B** Percentage yield = (3.00 g ÷ 4.61 g) × 100% = 65.1% **C** 13.8 is 10% of 138 (or 10.2 is 10% of 102), so only 10% of 180 g of aspirin can be made: 180 g ÷ 10 = 18.0 g

Lesson plan

Reactivate knowledge

Lesson resources

Activity: *Calculating percentage yield* (student sheet, teacher and technician sheet)

1. Which elements have the symbols N and I?
2. What is the chemical symbol of lead?
3. What is the relative mass of $C_2H_4O_2$? Use data from C3 4.7 *Relative mass*.

Answers: 1 Nitrogen and iodine **2** Pb **3** (2 × 12) + (4 × 1) + (2 × 16) = 60

Trigger interest

Show the image in Figure 1 of the Student Book of the reaction between lead nitrate and potassium iodide to produce the yellow solid, lead iodide. Ask students to recall how to separate an insoluble solid from a solvent (filtration, then drying accept decanting). Ask students how they could predict how much product we get (use the chemical equation and relative masses). Then ask students whether or not we always produce the amount of a product we were expecting (no) and why this might be (students may suggest: spillages, loss of reactants during transfer/getting left behind, incomplete reactions).

Exposition of main content

Introduction to percentage yield Show the word and balanced symbol equations for the production of lead iodide on the board with the reacting masses underneath. Ask students what they think the terms *theoretical yield*, *actual yield*, and *percentage yield* mean and where they may have heard them before.

Explain the term *theoretical yield* (the maximum possible mass of a product produced), and that the theoretical yield of lead iodide when 3.31 g of lead nitrate is used would be 4.61 g. Repeat the calculations of theoretical yield with different masses of reactants to check for misconceptions. Explain the term *actual yield* (the actual amount of a product made in a reaction), and explain that it may be that when this reaction was completed, only 4.15 g was produced. This is the actual yield.

Show the equation for calculating percentage yield on the board, explain the meaning of the term *percentage yield* (the percentage of the theoretical yield that is actually made in a reaction), and substitute in the values used earlier, to calculate the percentage yield for that reaction. Briefly discuss reasons why companies want high percentage yields (maximising profits from selling products whilst minimising waste and cost of reactants).

Some students may have the misconception that calculating the percentage yield involves the relative mass of the reactants. Although this may be needed to work out theoretical yields, it is only the theoretical yield of the products needed to work out the percentage yield.

Main activity

Activity: Calculating percentage yield (30–35 mins) Students complete the activity to practise calculating the theoretical yield and percentage yield.

Review and reflect

Use formative assessment and targeted questioning to ensure students feel confident in the differences between theoretical, actual, and percentage yields and how to calculate them, as well as why it is important for companies to monitor and be aware of them.

Homework

Students write instructions on how to calculate the theoretical yield from a balanced chemical equation, and how to calculate the percentage yield using this and the actual mass of a product. In future lessons, students could peer assess how clear each other's instructions are.

C3 Chapter 4: Useful chemical reactions

Introduction to checkpoint intervention

This checkpoint intervention provides suggestions for a lesson to follow up the C3 4 *Useful chemical reactions* checkpoint assessment. Depending on the outcomes of the assessment, these suggestions could move a class towards secure, to target specific learning outcomes, or to consolidate knowledge for students achieving secure outcomes. Use students' outcomes from the checkpoint assessment to address general misconceptions from the content covered. Be prepared to re-cover content as required; three key concepts from the chapter are covered in more detail below.

Checkpoint secure learning outcomes

- Explain how metal properties make them suitable for their uses.
- Determine the position of a metal in the reactivity series.
- Predict whether or not pairs of substances react in displacement reactions.
- Explain why different methods are used to extract different metals from their ores.
- Explain how catalytic converters make exhaust products less harmful.
- Calculate relative masses of molecular elements and compounds.
- Calculate percentage yields.

Student reflection

Encourage students to reflect on whether there were any checkpoint questions they found difficult or straightforward, as well as their preparation for the checkpoint assessment, including how they revised, the time spent on revision, and what areas they could improve on. For more strategies, see the Metacognition chapter or the Metacognition in Key Stage 3 Science guide on Kerboodle.

Intervention activity: Explain how metal properties make them useful

The most significant barrier to a firm understanding of the properties of metals is vocabulary. Some words will have been encountered before (e.g., conductor) but students may not realise that a material can be a thermal or an electrical conductor. Some words (e.g., malleable, ductile, sonorous) are likely to be new to students whereas others (e.g., dense and even shiny) may not appear 'scientific' enough for students to realise they can use them. Regularly assess students' scientific vocabulary.

This topic is an excellent opportunity for students to do some of their own research using the internet or the school library. Give each student a different metal and ask them to describe and explain their properties and applications. You can control the amount of stretch or support by which metal you give each student. Some examples are given in the following table.

Simpler examples	Intermediate examples	Harder examples
copper	palladium	osmium
iron	chromium	cadmium
aluminium	titanium	mercury
gold	lead	tungsten

You could also put as many elements as there are students in a pot and students pick one out at random. It is important to encourage students to make correct use of scientific vocabulary.

Students may work in groups or individually and may present their findings as a piece of extended writing, a poster, or a presentation to the class.

Intervention activity: Explain how catalytic converters make exhaust products less harmful

Students may have encountered catalysts before in the context of enzymes and digestion. If they have, explicitly make the link for them between this topic and enzymes.

With regard to the application of catalysis to catalytic converters, it is important for students to understand that combustion does not always produce only carbon dioxide and water. Not only is there incomplete combustion, leading to the formation of carbon monoxide, but also nitrogen in the atmosphere can form oxides of nitrogen (NO_x); both of these are not intentional but are harmful by-products. There are three stages to how catalytic converters work: first, molecules stick to the surface (adsorption), then the atoms rearrange themselves (reaction), and then the products leave the surface (desorption). It is important to emphasise that at the end of the process, the catalyst hasn't changed, meaning it can be used again.

Model an answer for the following question:

Explain how catalytic convertors remove carbon monoxide from exhaust fumes. **(4 marks)**

Students could then answer a similar question, for example, *Explain how catalytic convertors remove nitrogen monoxide from exhaust fumes.* **(4 marks)**

If you wish, you could ask students to explain how CO or NO were produced in the first place. **(2 marks)**

Intervention activity: Calculate percentage yields

At GCSE, students will learn about the mole concept, which helps work out theoretical yields. At this stage, it is enough for them to recognise that the ratio of relative formula mass (M_r) is the same as the ratio of masses of that chemical. The lead iodide example on C3 4.8 *Calculating yield* provides a good example. It is recommended to start with masses that are factors of 2, 10, 100, or 1000 (and variations thereof) of the relative formula mass. For example:

Chemical equation	$2C_2H_5OH$	+	$6O_2$	→	$4CO_2$	+	$6H_2O$
Relative formula mass	92		192		176		108
Example 1 (÷ 100)	0.92 g						
Example 2 (÷ 2)	46 g						
Example 3 (÷ 20)			9.6 g				

You can then repeat the example, but tell students how much of one of the products is made and ask them to work out a percentage yield. Students will first need to work out the theoretical yield. For example:

Chemical equation	$2C_2H_5OH$	+	$6O_2$	→	$4CO_2$	+	$6H_2O$	**Percentage yield**
Relative formula mass	92		192		176		108	–
Example 1 (÷ 10)	9.2 g		–		–		9.6 g	
Example 2 (÷ 200)	0.46 g		–		0.66 g		–	
Example 3 (÷ 1000)	0.092 g		–		–		0.1 g	

3 Physics

Introduction to unit

This unit builds on all the key themes covered in KS3, extending familiar ideas to more challenging situations and taking an increasingly quantitative approach to physics concepts. Links are made to many important applications of these concepts, including medical applications and electricity generation. There are opportunities for discussion of the public understanding of science and science literacy built into the unit. Investigative skills are also developed, including taking a more quantitative approach to the analysis of results.

Physics NC links

- Energy
- Motion and forces
- Waves
- Electricity and electromagnetism

Working scientifically links

Each lesson lists the relevant Working scientifically National Curriculum links at the top of the spread. For further details, please refer to the *Oxford Smart Curriculum for Science* document.

Learning journey

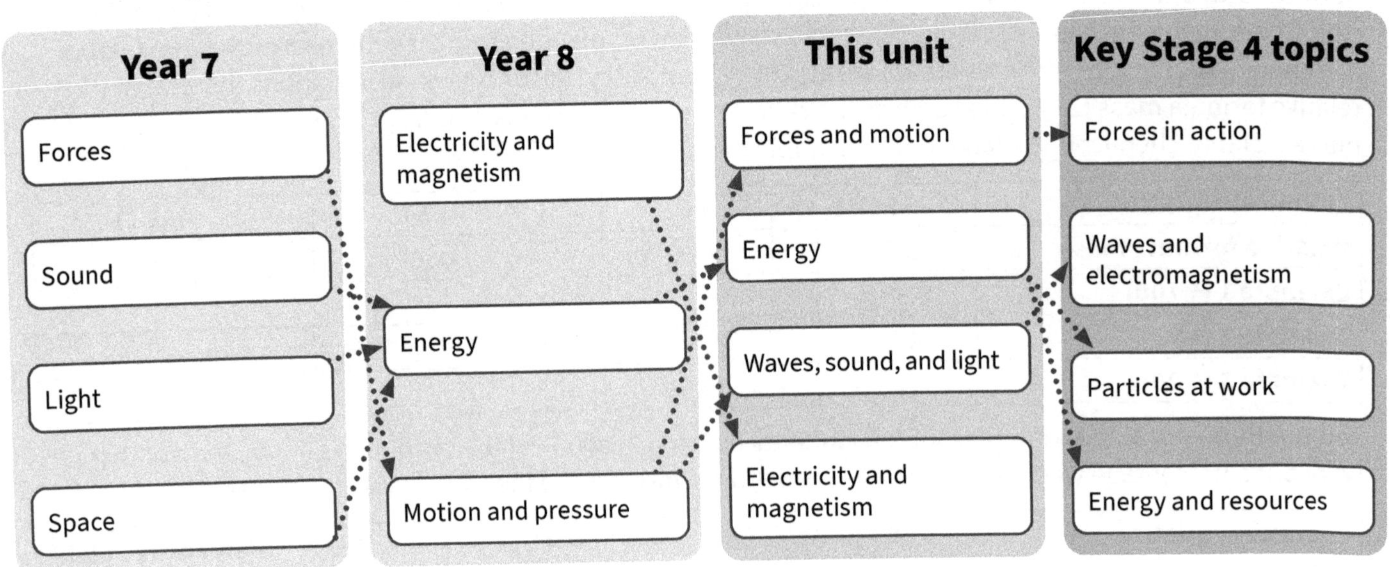

Physics and you

Physics underpins the technologies that shape our lives. From cars and planes to computers and mobile phones, washing machines and microwave ovens to ultrasound scans and radiotherapy, modern technologies allow us to do things people could only dream of 200 years ago. Invite students to identify these technologies and imagine how their lives would be different without them. You might like to divide the class into groups and give each group a setting or area of life, such as hospitals, transport, or entertainment. You could link technological development to social change, for example, the changes to people's lives made possible by the invention of the washing machine. Students might not know that before washing machines, a person could spend a whole day every week washing clothes!

Physics and the world

The challenges of climate change and, more recently, the cost of electricity and fuel are in the news on a regular basis. Students may have opinions and stories to tell related to these challenges. Though technology alone cannot deliver a 'magic solution', technologies that reduce the amount of energy we use and especially how much fossil fuel we burn have a part to play in meeting these challenges. Students may be familiar with examples of more efficient appliances, such as energy-saving light bulbs, and ways of generating electricity from renewable resources, such as solar panels, which show how physics knowledge can contribute via technological developments. Electric cars are becoming more common and students may know someone who owns one. Students can be invited to imagine future technologies that will help make our lifestyles more sustainable and also the limits of this approach.

The importance of communicating scientific and technological issues to the public is also of great importance. Students may have heard that mobile phone masts were vandalised during lockdown by people who believed the masts spread COVID-19. Journalists and presenters with a good understanding of physics can better communicate potential risks and benefits. Students can be invited to discuss what might cause someone to believe a mobile phone mast spreads COVID-19 and how, as a society, we might avoid people believing ideas that scientists know to be false.

Big questions

How will we travel in the future? Hydrogen fuel cells are being developed as an alternative to powering vehicles with petrol and diesel. For these to reduce carbon dioxide emissions, the hydrogen needs to be made using energy from renewable resources. Likewise, electric cars reduce emissions if the electricity they use has been generated from a low emission, renewable source.

How will we charge our devices in the future? Students may already have seen wireless charging devices. Battery technology continues to improve, allowing smaller and smaller batteries to store the same amount of energy, which in turn allows electronic devices to be made smaller and smaller. You can show students pictures of mobile phones through the years to illustrate this. You can invite students to think about what this means for drone technology.

Where can this take you? As well as jobs in structural, mechanical, and electrical engineering, a background in physics is also a pathway to jobs in medicine, research, patent law, the food industry, teaching, publishing, science journalism, and others, as well as careers that require the mathematical skills that physicists have, such as banking.

P3 Chapter 1: Forces and motion

Introduction to chapter

This chapter applies concepts students have learnt in KS3, such as balanced and unbalanced forces, to more challenging situations, such as a skydiver. In KS3, students have begun to quantify motion and physical properties, such as the stiffness of a spring, and have calculated speed and weight. This chapter extends quantification to resultant force and acceleration. Distance–time graphs are reviewed and speed–time graphs introduced. Instantaneous speeds and accelerations are calculated from tangents.

Core concepts

- Forces: types and interaction pairs
- Force diagrams
- Motion
- Stretching and Hooke's Law
- Forces at a distance
- Mass and weight
- Unbalanced forces
- Speed

Prerequisite knowledge

- Representing forces using arrows
- How contact forces arise as a result of interactions at a microscopic level
- The idea of a 'field'
- The difference between weight and mass, and how to calculate weight
- Quantifying the stiffness of a spring using Hooke's Law
- The idea that forces can act in the same or in opposite directions, and that resultant force can change an object's speed and/or direction
- Calculating speed
- Using distance–time graphs to 'tell a story'; the idea of 'rate' being related to slope

GCSE links

- Forces as vectors
- Representing equilibrium and non-equilibrium situations using free-body diagrams
- Acceleration caused by forces
- Newton's First and Third Laws
- Speed of sound, estimating speeds and accelerations in everyday contexts
- Interpreting quantitatively graphs of distance, time, and speed
- Explaining the shape of displacement–time and velocity–time graphs; calculating speed and acceleration using tangents of a graph

Key learning misconceptions

- Gravity only affects heavy things and stops acting when the object hits the ground.
- Gravity only works one way – the Earth attracts the Moon, but the Moon does not attract the Earth.
- Mass and weight mean the same thing.
- Astronauts are 'weightless' in an orbiting spacecraft because there is no gravity.
- Heavier objects always fall faster than lighter objects.
- If an object is stationary, there are no forces acting on it, and if it is moving there must be a force acting on it.
- Acceleration can only occur in the same direction as an object is moving.
- Friction only occurs between solid objects and when an object is moving.
- Opening a parachute during free fall, when gravity is the only force acting on the skydiver, makes the skydiver go upwards.
- Pressure and force mean the same thing.
- Objects float in water because they contain air.
- The weight of an object resting on a surface and the normal reaction of the surface are an interaction pair.

Teaching preparation

Teaching strategy

Students often find it hard to identify interaction pairs because they are not used to thinking about, for example, the weight force they exert on the Earth. It is worth spending some time identifying examples as a class; no example is too simple! Remember that the two forces in an interaction pair will be the same size and same type of force, but will act in opposite directions and on 'the other object'. For example, if the Earth exerts a gravitational force downwards on me, I must be exerting a gravitational force upwards on the Earth.

Key maths skill

Units As students start to look at data quantitatively, it is important to emphasise that you can *only put numbers into a formula if the quantities have the right units*. Converting from miles per hour to metres per second is useful, but challenging for students.

Key maths skill

Gradients of straight lines The weight formula provides a chance to introduce $y = mx$ graphs in a physics context (P3 1.3 *Mass, weight, and fields*). With weight on the y-axis and mass on the x-axis, the gradient is the gravitational field strength of the planet/moon. Because the field strength on the surface of a planet/moon is just a number (i.e., a constant), we know that the graph will be a straight line through the origin. Students may not have covered the equations of straight lines in maths lessons yet, so be prepared to teach this maths skill from scratch.

Practical demonstration

P3 1.5 *Balanced and unbalanced* forces Students will be impressed if you can whirl a tray of cups of water round in a vertical circle! And it is not as difficult as it might appear, so it is worth mastering before the lesson. It is also safe enough for students to have a go themselves, making for a memorable lesson.

Key maths skill

Tangents There is a good chance your students will not know what a tangent is or how to draw one, so this should be approached as though you are introducing it for the first time (P3 1.4 *Speed and distance–time graphs*). Model drawing tangents by putting your ruler on the board and showing students how to position it.

Key learning misconception

Constant acceleration and no acceleration The concept of constant acceleration can be challenging for students. An object with constant acceleration moves further each second than it did the second before. If students have not distinguished clearly between acceleration and speed, this can be very confusing.

The explanation of how a skydiver reaches terminal velocity is particularly difficult for students to master, partly due to the misconception that, if the forces on you are balanced, you are stationary. Repeat often that if the forces on you are balanced, you are *not accelerating*.

Key learning misconception

Centrifugal force Students may have heard of 'centrifugal force' and think that it is, for example, the force that throws you outwards when you go round a tight corner. In fact, nothing is throwing you outwards; you are trying to continue in a straight line, but the vehicle you are travelling in is accelerating you towards the centre of the corner. Consequently, centrifugal force is known as a pseudo-force; it is not a real force, but appears to be a real force to us in our everyday experience. This can be a nice opportunity to talk about counter-intuitive conclusions.

1.1 Reactivate: Forces and motion

Themes: Forces and motion

Sub-themes: Balanced and unbalanced forces, Newton's Laws

Physics NC links

- opposing forces and equilibrium: weight held by stretched spring or supported on a compressed surface
- forces being needed to cause objects to stop or start moving, or to change their speed or direction of motion (qualitative only)

Learning objective
• To recall and organise key knowledge about forces and motion.

Year 7 learning objectives to revise

- Use force arrows to explain how forces arise
- Describe the difference between balanced and unbalanced forces
- Explain why an object is in equilibrium
- Explain the changing motion of objects

Year 8 learning objectives to revise

- Give the definition of an electric field
- Describe how to investigate and represent the shape of a magnetic field
- Calculate average speed
- Describe the difference between average and instantaneous speed
- Describe motion using a distance–time graphs
- Use data from a distance–time graph to calculate average speed
- Give the definition of atmospheric pressure
- Give the definition of the moment of a force

Pinchpoint focus

Motion of objects in equilibrium.

Tier 2 vocabulary	Tier 3 vocabulary
acceleration, equilibrium, weight	Newton's First Law, terminal velocity, velocity

Trigger interest

Show students the image of a person in a jet suit from the Lesson presentation. Ask: *Can you use forces to explain how this person is hovering?*

Lead discussion to the fact that the forces are balanced.

(Note: The company that makes the jet suit – *Gravity* – has many exciting online videos that you could show.)

Drop identical ball bearings into two tall measuring cylinders, one filled with water and one with a viscous (but transparent) liquid, such as bubble bath. Ask why the ball bearing in bubble bath falls at a constant speed while the ball bearing in water speeds up.

Release a ping pong ball and a metal ball bearing at the surface of water in a beaker. Ask why the ping pong ball floats on the surface, but the metal ball bearing accelerates downwards.

Recall

Give students the short quiz revising forces and equilibrium on the Lesson presentation.

If students finish quickly, ask them to use key terms (such as resultant force) to write sentences about the person in the jet suit.

Recap how to draw free-body diagrams; draw them for the situations outlined in Trigger interest.

Main activity

Pinchpoint: Forces and motion (20–25 mins)

The pinchpoint question is about a car travelling on a straight, flat, road. It tests understanding of equilibrium, which is fundamental to understanding object motion.

Students answer the pinchpoint question and complete the follow-up activity with the same letter. They then complete the *What I need to remember* activity. Information about each incorrect answer's misconceptions can be found on the teacher sheet. You may want to divide the class into groups, depending on the letter answer chosen, to support each other when completing the follow-up activity.

Students measure the terminal velocity reached by a ball bearing in bubble bath. They release the ball bearing in the bubble bath in a measuring cylinder and mark the position of the ball every second. When the distances between the marks are the same, the ball has reached terminal velocity.

Students use two light gates and a trolley on a track to measure the trolley's speed when the track is:

- 'flat'
- slanted downhill
- slanted uphill.

Review and reflect

Students are shown the pinchpoint question again and asked to reflect on whether they think they chose the correct letter and if they would change their answer now. If they think they got it wrong, ask them to think about why they chose that answer.

Reveal the correct answer. Review with students why it is correct and the other options are not. Guidance can be found on the teacher sheet. Students could be asked to pick an incorrect answer and write an explanation to a fictional student who chose it, explaining why it cannot be correct.

Discuss why the speed of the ball bearing doesn't change during this time.

Discuss why the speed of the trolley does not change on the flat track.

Homework

Ask students to complete the knowledge organiser.

1.2 Apply: Forces and interactions

Theme: Forces and motion

Sub-themes: Balanced and unbalanced forces, Newton's Laws, Representing forces, Types of force

Physics NC links

- forces as pushes or pulls, arising from the interaction between two objects
- using force arrows in diagrams, adding forces in one dimension, balanced and unbalanced forces
- non-contact forces: gravity forces acting at a distance on Earth and in space, forces between magnets and forces due to static electricity
- forces being needed to cause objects to stop or start moving, or to change their speed or direction of motion (qualitative only)

Working scientifically NC links

- use appropriate techniques, apparatus, and materials during fieldwork and laboratory work, paying attention to health and safety
- present observations and data using appropriate methods, including tables and graphs
- interpret observations and data, including identifying patterns and using observations, measurements, and data to draw conclusions

Learning objective
Use Newton's Third Law to explain how forces arise, and change motion.

Year 7 and 8 learning objectives to be applied	Application
• Use force arrows to explain how forces arise • Describe the difference between balanced and unbalanced forces • Explain the changing motion of objects • Give the definition of atmospheric pressure	• Buoyancy, rockets, and thrust

Tier 2 vocabulary	Tier 3 vocabulary
interaction, magnitude	contact force, interaction pair, Newton's Third Law, non-contact force, upthrust

SB in-text question answers

A They are different types of force (weight is a non-contact force (of gravity) whereas the reaction is a contact force) **B** A balloon full of air would have greater weight than upthrust, which would stop it from floating **C** Force of the rocket on the Earth

Lesson plan

Reactivate knowledge

Lesson resources

Practical: *Upthrust* (student sheet, support sheet, teacher and technician sheet)

1. Name three contact and three non-contact forces.
2. Write down Newton's Third Law.
3. Describe how you show a force on a diagram.

Answers: 1 Non-contact forces: magnetism, gravity, electrostatic; contact forces: three from tension, compression, drag, friction, buoyancy **2** Forces come in pairs **3** With an arrow attached to the object/dot

Note**:** For Newton's Third Law, many students may say 'For every action there is an equal and opposite reaction'. We avoid this definition in favour of the one that talks explicitly about forces.

Trigger interest

Show students the three pictures on the Lesson presentation and ask: *What connects these three objects?* Guide discussion towards a definition of Newton's Third Law.

Exposition of main content

Show students the diagram of the newtonmeter on the Lesson presentation. (Note: you could hang a block from a newtonmeter to demonstrate this for the students.) Explain what is happening, in terms of gravitational force.

Ask: *What do you think will happen when you lower the block into the water?* Discuss answers and then show students the next diagram on the Lesson presentation. (Again, you could demonstrate lowering the block into a beaker of water.) Ask: *Why is the tension in the newtonmeter less now?* Discuss ideas before providing the answer.

Main activity

Practical: Upthrust (40 mins) Students observe and measure upthrust forces. They make the connection between the size of the upthrust and the amount of water displaced. Students measure blocks of different sizes, calculate their volumes, and measure their weights. They then submerge the blocks in water, calculate the upthrust, and plot a graph of upthrust against volume.

Review and reflect

Hold a ping pong ball underwater, then let go. Students explain why the ping pong ball accelerates upwards. (Upthrust force is greater than the weight. Once it is moving, there is a downwards drag force as well.)

Homework

Students look for an interesting example where there is an upthrust force. (Some possibilities: upthrust on helium balloon from air, people floating in the Dead Sea, liquids of different densities floating on top of each other, brine pools at the bottom of the sea.)

1.3 Apply: Mass, weight, and fields

Theme: Forces and motion

Sub-theme: Types of force

Physics NC links

- forces as pushes or pulls, arising from the interaction between two objects
- using force arrows in diagrams, adding forces in one dimension, balanced and unbalanced forces
- non-contact forces: gravity forces acting at a distance on Earth and in space, forces between magnets and forces due to static electricity
- gravity force, weight = mass × gravitational field strength (g), on Earth g = 10 N/kg, different on other planets and stars; gravity forces between Earth and Moon, and between Earth and Sun (qualitative only)

Working scientifically NC links

- pay attention to objectivity and concern for accuracy, precision, repeatability, and reproducibility
- present observations and data using appropriate methods, including tables and graphs
- interpret observations and data, including identifying patterns and using observations, measurements, and data to draw conclusions

Learning objectives

Describe the link between forces and fields, and do calculations.

Compare systematic and random errors.

Year 7 and 8 learning objectives to be applied

- Use force arrows to explain how forces arise
- Describe how to investigate and represent the shape of a magnetic field
- Give the definition of an electric field

Application

- Investigating how to measure gravitational field strength.

Tier 2 vocabulary	Tier 3 vocabulary
average, gradient, inertia, mass, weight	directly proportional, graduation, gravitational field, random error, systematic error

SB in-text question answers

A There is a gravitational field line between the Earth and the object pointing towards the Earth, so it experiences a force downwards **B** Systematic error **C** 3 N

Lesson plan

Reactivate knowledge

1 Draw a force diagram for a person in bed.

2 Write down the units of mass and weight.

3 Write down the equation for calculating weight.

Lesson resources

Practical: *Gravitational field strength on Planet X* (student sheet, support sheet, teacher and technician sheet), Homework: *Measuring gravitational field strength* (homework sheet)

Answers: 1 Dot with arrow down labelled the force of the Earth on the person; arrow upwards labelled force of the bed on the person **2** Mass – kg, weight – N **3** Weight = mass × gravitational field strength

Trigger interest

Ask: *Why would you lose weight if you went on holiday to the Moon?* Discuss the difference between mass and weight before showing the two definitions on the Lesson presentation.

Exposition of main content

Thinking about mass Discuss the two ways of thinking about mass, as shown on the Lesson presentation. Ensure students are familiar with the terms inertia, weight, force, and gravitational field.

Do you have gravity? Ask: *Objects accelerate downwards because of the Earth's gravity… but why don't they accelerate towards you, because of your gravity?*

What is *g*? Recap what g stands for in the formula for weight. Explain that the gradient of the graph on the Lesson presentation is equal to the gravitational field strength of the Earth.

Systematic or random? Ask students to discuss, in pairs, whether the errors described on the Lesson presentation are systematic or random. Discuss as a class.

During this section, you could try holding up a newtonmeter with a mass hanging off it, and ask: *What might cause random error in your weight measurement?* (Answer: Mass bouncing.)

Main activity

Practical: Gravitational field strength on Planet X (40 mins) Explain that students need to plan a practical to determine the gravitational field strength on a newly discovered planet. Work through the Lesson presentation to help them decide:

- how they will select the most appropriate measuring instrument
- what range they will choose
- how they will avoid systematic errors and minimise random error.

Allow students to carry out their practical, using the objects labelled with fake masses.

Review and reflect

Ask students to justify the choices they made when planning their practical.

Homework

Students are given a list of six measuring instruments. For each one, they choose an example of something that it could be used to measure.

1.4 Apply: Speed and distance–time graphs

Theme: Forces and motion

Sub-theme: Motion

Physics NC links

- speed and the quantitative relationship between average speed, distance, and time (speed = distance ÷ time) (KS3)
- the representation of a journey on a distance–time graph (KS3)
- speed of sound, estimating speeds and accelerations in everyday contexts (KS4)
- interpreting quantitatively graphs of distance, time, and speed (KS4)

Working scientifically NC link

- apply mathematical concepts and calculate results

Learning objective

Make estimates about everyday speed, and use a distance–time graph to calculate speed.

Year 7 and 8 learning objectives to be applied

- Calculate average speed
- Describe the difference between average and instantaneous speed
- Describe motion using a distance–time graph
- Use data from a distance–time graph to calculate average speed

Application

- Estimating and calculating the speeds of animals.

Tier 2 vocabulary	Tier 3 vocabulary
accelerate, average, estimate, speed, tangent	convert unit, distance–time graph

SB in-text question answers

A This speed is the constant speed required to cover the total distance in the total time and it need not match the speed at a particular instant
B 1609 × 15 = 24 135 m

Lesson plan

Reactivate knowledge

Lesson resources

Activity: *Calculating speed* (student sheet, support sheet, teacher and technician sheet)

1. What two quantities do you need to calculate speed?
2. Write down the units of speed in science.
3. What does a steep slope on a distance–time graph mean?

Answers: **1** Distance and time **2** Metres per second **3** Fast speed

You could also ask what it means if a line is getting steeper or less steep.

Trigger interest

Ask: *What were the highest and lowest speeds that you achieved yesterday?* Make the point that 0 m/s is the lowest speed.

Exposition of main content

Average speed Recap the average speed formula by asking which equation links distance (s), speed (v) and time (t). Explain why estimating is useful. Ask students to estimate some everyday speeds, such as walking and cycling.

Converting m/s to mph Explain how to convert between m/s and mph. Use the example on the Lesson presentation. You could use other examples, such as the highest or lowest speeds given in the Trigger interest discussion.

Distance–time graph gradients Recap how to find the gradient of a straight line (change in y divided by change in x) and that speed is the gradient of a distance–time graph. Point out that a straight line indicates constant speed, an upwards curving line indicates that the object is getting faster, and a downwards curving line indicates that the object is getting slower.

Demonstrate drawing a tangent and finding its gradient for a point on a curve.

Main activity

Activity: Calculating speed (30 mins) Students estimate the speed of various animals, use the data to calculate average speeds, and complete questions by interpreting a distance–time graph.

Review and reflect

Ask students to peer assess their results from the Lesson presentation slides. Are the students surprised by their answers?

Invite students to walk across the classroom, varying their speed, while their classmates attempt to draw distance–time graphs for their journeys on mini-whiteboards.

Homework

Ask students to research race times for three different races, such as a marathon, the Tour de France cycling race, or the final of the mixed relay swim in the 2021 Olympics. Ask them to calculate the average speed of the competitors.

1.5 Apply: Balanced and unbalanced

Theme: Forces and motion

Sub-themes: Balanced and unbalanced forces, Newton's Laws

Physics NC links

- using force arrows in diagrams, adding forces in one dimension, balanced and unbalanced forces
- opposing forces and equilibrium: weight held by stretched spring or supported on a compressed surface
- forces being needed to cause objects to stop or start moving, or to change their speed or direction of motion (qualitative only)
- change depending on direction of force and its size

Learning objective

Explain equilibrium and circular motion in terms of forces.

Year 7 and 8 learning objectives to be applied

- Describe the difference between balanced and unbalanced forces
- Explain why an object is in equilibrium
- Explain the changing motion of objects
- Give the definition of a moment of a force

Application

- Acceleration caused by forces; Newton's First Law

Tier 2 vocabulary	Tier 3 vocabulary
equilibrium, gradient, inertia, weight	centripetal force, Newton's First Law, terminal velocity, velocity

SB in-text question answers

A The astronaut's toolkit is in equilibrium since its motion does not change **B** C and E **C** The direction of motion keeps changing, so the motion is not constant

Lesson plan

Reactivate knowledge

Lesson resources

Activity: *Surface area and drag* (student sheet, support sheet, teacher and technician sheet)

1. Define balanced in terms of forces.
2. Define unbalanced in terms of forces.
3. Name two drag forces.

Answers: 1 Balanced means the forces cancel out **2** Unbalanced means the forces do not cancel out **3** Air resistance, water resistance

Trigger interest

Ask students if they agree with the statement on the Lesson presentation. Facilitate a class discussion; include these points: Who do we trust and why? The idea is quite intuitive; if I stop pushing something, it stops moving. How do long-held ideas get overturned?

Exposition of main content

Distance–time graph for a skydiver Show the image on the Lesson presentation. Discuss the changing forces on a skydiver. Emphasise that the drag force gets bigger as the skydiver gets faster.

Falling paper demonstration Drop a large paper cone onto the floor, from a height. Then scrunch the paper up into a ball and drop it again from the same height. Ask students what they notice about the speeds at which the paper shapes have fallen. Point out that the cone reaches its terminal velocity very quickly and floats down slowly, due to its large surface area.

Main activity

Activity: Surface area and drag (25 mins)
Part 1: Ask students to work through the two scenarios on the Lesson presentation. The activity sheet has space for their answers.

Part 2: Carry out the 'swinging bucket' demonstration as described in the teacher notes. Then ask students to complete the questions on the activity sheet.

Review and reflect

Discuss the statements on the Lesson presentation. Do students agree or disagree? If they disagree, what would they say instead?

Homework

Ask students to write down examples of:

- an object in equilibrium (forces balanced)
- an object getting faster or slower (forces unbalanced)
- an object moving at a constant speed in a circle (forces unbalanced)

and identify the forces acting and draw a force diagram for each object.

1.6 Reach: Resultant forces

Theme: Forces and motion

Sub-themes: Balanced and unbalanced forces, Representing forces

Physics NC links

- opposing forces and equilibrium: weight held by stretched spring or supported on a compressed surface (KS3)
- forces being needed to cause objects to stop or start moving, or to change their speed or direction of motion (qualitative only) (KS3)
- change depending on direction of force and its size (KS3)
- forces as vectors (KS4)

Learning objective
Use ideas about vectors and scalars to calculate resultant forces.

Year 7 and 8 learning objectives to be built on
• Use force arrows to explain how forces arise • Describe the difference between balanced and unbalanced forces • Explain why an object is in equilibrium • Explain the changing motion of objects

Key Stage 4 link
• Forces as vectors

Tier 2 vocabulary	Tier 3 vocabulary
displacement	resultant force, scalar, vector

SB in-text question answers
A Tension, weight **B** 0 N **C** 0 m

Lesson plan

Reactivate knowledge

1 Write down what an object does when the forces are balanced.
2 Write down what an object does when the forces are unbalanced.
3 Name the force that keeps the Moon in orbit.

Lesson resources
Activity: *Forces carousel* (student sheet, support sheet, teacher and technician sheet), Homework: *Resultant forces* (homework sheet)

Answers: **1** Stay still, move with a constant speed **2** Accelerate, speed up, slow down, change direction **3** Gravity

Trigger interest

Show students the image of the tug of war in the Lesson presentation. Ask: *What will happen if one side lets go?* Draw out the idea that the motion of the rope is determined by the sum of the forces on it.

Exposition of main content

Distance and displacement Ask students the questions on the Lesson presentation and discuss their answers. Ask: *In each example, how far did you travel? What was your displacement?* Guide discussion to the concept of scalar and vector quantities. Ask: *What affects total distance? What affects total displacement?* Discuss answers. Ask: *Can you think of any other vector quantities?* Guide discussion until someone suggests 'force'.

Force diagrams Show students the force diagrams on the Lesson presentation. Ask: *Are these forces balanced? How can you tell?* Explain what a resultant force is and how to calculate the size and direction of resultant forces. Emphasise that you choose a direction to be positive and then forces in the other direction are negative.

Main activity

Activity: Forces carousel (30 mins) Students move around the classroom to different activity stations, each with an object. At each station, they follow simple instructions, then draw a force diagram and decide if the object is in equilibrium or not. They then complete questions on the activity sheet.

Review and reflect

Students compare their answers for each activity to those on the Lesson presentation.

Use the picture of the rocket just after launch as a prompt to review the key ideas from the lesson. At the moment of launch, the thrust force upwards on the rocket produced by the engines is greater than the weight force downwards exerted by the Earth, so there is a resultant force upwards. That is why the rocket accelerates upwards.

(Students who have worked on the extension task may be able to add that once the rocket is moving, there is an air resistance force downwards, reducing the resultant force. This would reduce the acceleration. However, they may also realise that the mass of the rocket is decreasing as the fuel is used up, which would increase the acceleration. We do not know which would have the bigger effect.)

Homework

Resultant force Students draw the path of a journey on a grid from a list of instructions. They work out the total distance of the journey and the total displacement from the starting point.

1.7 Reach: Acceleration and speed–time graphs

Theme: Forces and motion

Sub-theme: Motion

Physics NC links

- speed of sound, estimating speeds and accelerations in everyday contexts (KS4)
- interpreting quantitatively graphs of distance, time, and speed (KS4)

Working scientifically NC link

- translating data from one form to another

Learning objective

Calculate acceleration, and deduce types of motion from a speed–time graph.

Year 7 and 8 learning objectives to be built on

- Calculate average speed
- Describe the difference between average and instantaneous speed
- Describe motion using a distance–time graph
- Use data from a distance–time graph to calculate average speed

Key Stage 4 links

- Speed of sound, estimating speeds and accelerations in everyday contexts
- Interpreting quantitatively graphs of distance, time, and speed

Tier 2 vocabulary	Tier 3 vocabulary
acceleration, gradient, rate, tangent	

SB in-text question answers

A – **B** 0 m/s^2

Lesson plan

Reactivate knowledge

1 Write the equation for speed.

2 Describe the link between speed and the slope of a distance–time graph.

3 Write down how to find the slope at a point on a curved line.

Lesson resources

Activity: *Speed–time graphs* (student sheet, support sheet, teacher and technician sheet)

Answers: 1 Speed = distance/time **2** The steeper the slope, the bigger the speed **3** Draw a tangent to the curve at that point; find the slope of the tangent

Trigger interest

Ask students: *Which organism can produce the greatest acceleration?* Discuss their ideas, before revealing the answer: a fungus called *Pilobolus crystallinus* (the fungus cannon).

(Note: you could play a BBC Earth Unplugged video that shows the fungus cannon 'firing' in slow motion, although care should be taken to avoid content about guns.)

Exposition of main content

Dropping objects in a vacuum Show students the diagram in the Lesson presentation. Explain what 'rate' and 'acceleration' (the amount by which speed increases in one second) mean. Remind them that acceleration due to gravity is $10\,m/s^2$. Ask: *Would the dropped objects have more or less acceleration if there were not a vacuum? Why?*

Calculating acceleration Remind students how to calculate acceleration. Work through the examples on the Lesson presentation.

Speed–time graphs Use the Lesson presentation to show what different accelerations look like on a speed–time graph. Recap how to find the gradient of a curve. You could ask a volunteer to act out the motion shown on the graph(s) by walking at varying speeds as you call out the time.

Main activity

Activity: Speed–time graphs (30 mins) Students complete questions based on interpreting a speed–time graph. They then answer a six-mark question shown on the Lesson presentation. This asks them to describe the acceleration and speed of an object during the journey shown in the speed–time graph.

Review and reflect

Students peer-mark each other's answers. Then they answer the review questions shown on the Lesson presentation on mini-whiteboards.

Homework

Students find out the acceleration of their favourite car (or animal) in mph/s (e.g., 0–60 mph in 3 seconds = 20 mph/s on average), and convert the value to m/s^2.

Note: you may need to remind them how to convert from mph to m/s.

P3 Chapter 1: Forces and motion

Introduction to checkpoint intervention

This checkpoint intervention provides suggestions for a lesson to follow up the P3 1 *Forces and motion* checkpoint assessment. Depending on the outcomes of the assessment, these suggestions could move a class towards secure, to target specific learning outcomes, or to consolidate knowledge for students achieving secure outcomes. Use students' outcomes from the checkpoint assessment to address general misconceptions from the content covered. Be prepared to re-cover content as required; three key concepts from the chapter are covered in more detail below.

Checkpoint secure learning outcomes

- Use Newton's Third Law to explain how forces arise, and change motion.
- Describe the link between forces and fields, and do calculations.
- Compare systematic and random errors.
- Make estimates about everyday speed, and use a distance–time graph to calculate speed.
- Explain equilibrium and circular motion in terms of forces.

Student reflection

Encourage students to reflect on whether there were any checkpoint questions they found difficult or straightforward, as well as their preparation for the checkpoint assessment, including how they revised, the time spent on revision, and what areas they could improve on. For more strategies, see the Metacognition chapter or the Metacognition in Key Stage 3 Science guide on Kerboodle.

Intervention activity: Use Newton's Third Law to explain how forces arise, and change motion

Students often don't appreciate that forces are exerted by objects on other objects. In other words, you can't have a force 'on its own'; there have to be two objects involved. It is the forces exerted by these two objects on each other that form a force interaction pair.

Interaction pairs

Set up a series of objects (such as a ping pong ball floating in water, or a mass hanging from string). Ask students to identify as many force interaction pairs as they can in each situation.

Intervention activity: Make estimates about everyday speed and use a distance–time graph to calculate speed

Students often forget that *average velocity* is not the same as *instantaneous velocity*. Instead of drawing a tangent to the distance–time curve at a point, students often draw a straight line from the point to the origin.

Curved distance–time graphs

Practise finding speeds from curved distance–time graphs. You can create your own, or find examples online, for example, an Internet search for 'Usain Bolt graph' should find a distance–time graph and a speed–time graph for Usain Bolt's world record 100 m sprint at the 2009 World Championships in Berlin.

Intervention activity: Explain equilibrium and circular motion in terms of forces

Most students understand the concept that a stationary object is in equilibrium. However, many find it counterintuitive that an object moving with constant velocity is also in equilibrium. Remind students that resultant force causes *acceleration*, not motion. A special case is motion in a circle, where the acceleration is only changing the direction; the object doesn't speed up or slow down.

Vacuum tube

Find a large, see-through cylinder that you can evacuate with a vacuum pump. Put a penny and a flat piece of paper in the tube. Demonstrate that both accelerate (and fall at the same rate), when the tube is evacuated and there is no air resistance. Then allow air back into the tube and repeat.

The penny's weight is much greater than the air resistance on it, so it still accelerates. The paper's weight is very small and so the air resistance on the paper does not grow to the same size as the weight. The paper falls at a constant (terminal) velocity, even though it has forces acting on it, because those forces are balanced.

P3 Chapter 2: Energy

Introduction to chapter

In this chapter, students engage more fully with the fundamental idea of energy stores and begin to quantify how much energy is in them and moving between them. Examples of energy stores include the thermal store of a ball or the air in a room and the kinetic store of a moving object. Energy is an abstract concept that can be stored in any of these stores and can move between them, but cannot be created or destroyed. When energy moves from one store to another, something changes in the physical world, for example, the temperature of an object goes down or a moving object gets faster.

They will also consider factors that affect how quickly energy is transferred and how much is dissipated. When energy moves from one store to another, generally some of it is dissipated. This means that instead of going into the store you want it to go into, it goes into thermal stores, for example, the thermal store of the surrounding air. This is why dissipated energy is usually called wasted energy. While students generally find the idea of dissipation accessible, they find efficiency a more challenging concept and may have difficulty identifying which store contains the usefully transferred energy.

Prerequisite knowledge

- Energy is quantifiable
- Energy is stored in different types of store and can be transferred between them
- Energy can transfer by heating in the following ways: conduction, convection, radiation
- Infrared radiation is a wave, like (visible) light, that can travel through a vacuum
- Energy cannot be created or destroyed
- Work done is energy transferred, work done formula
- Machines make jobs easier, but the total amount of work done is the same
- Power is the rate of transfer of energy, the power formula

Core concepts

- Energy conservation, transfer, and dissipation

GCSE links

- Specific heat capacity and specific latent heat
- Gravitational energy formula
- Kinetic energy formula
- Elastic energy formula
- Efficiency formula

Key learning misconceptions

- Energy is a tangible thing.
- 'Heat' is a substance which can flow from place to place.
- An object at rest has no energy.
- Energy and force are interchangeable terms.
- Energy gets used up/energy can be made.
- Devices use up energy.
- Energy is confined to some particular origin, such as food, electric companies.
- 'Heat' rises (as opposed to hot substances rising).
- Thermal conductors and insulators are opposites, not part of a continuum.
- Hot objects can cool down without something else around them getting hot.
- Cold can be transferred.
- When you heat a substance, particles get hotter.
- Energy is fuel.
- Work is synonymous with labour or a job, and not to do with forces.

- Power and energy are the same thing.
- Heat and temperature are the same thing.
- Heating water that is at boiling point will make its temperature rise above 100 °C.
- Boiling point is the maximum temperature a substance can reach.
- The bubbles in boiling water contain air/oxygen or nothing.
- Some objects are naturally warmer than others.
- There are energy types that are distinct from each other.
- The only type of potential energy is gravitational.
- Gravitational potential energy depends only on the height of an object.
- All renewable energy resources are good, and all non-renewable energy resources are bad.
- All power stations produce smoke.
- Wind generators are very loud.

Teaching preparation

Modelling

Modelling energy stores with labelled jugs and pouring water between them is an effective way to model energy transfers and conservation of energy/dissipation.

Causes of confusion

Energy conservation versus wasted energy Some students may be confused by how energy appears to disappear when it dissipates. It is, therefore, important to make clear the energy stores that the dissipated energy has transferred into.

Sticking point

Work done Many students find it easy to talk about energy transferred, but confusing to talk about work done. It is, therefore, necessary to stress the equivalence often.

Key maths skill

Manipulating formulae Always model how to use and rearrange formulae. Note the units required and how to convert, if the context requires conversions. The elastic energy formula has an x^2 in it; it is a good idea to check students know how to handle this.

Working scientifically and literacy link

Weighing arguments Students will have the opportunity to role-play government ministers and scientific advisors, using evidence to decide whether the government should invest in solar roads. You may wish to decide in advance which students will take which roles and who they will work with.

Theme: Energy

Sub-theme: Energy transfer

Physics NC link

- heating and thermal equilibrium: temperature difference between two objects leading to energy transfer from the hotter to the cooler one, through contact (conduction) or radiation; such transfers tending to reduce the temperature difference: use of insulators

Learning objective
• To recall and organise key knowledge about energy.

Year 7 learning objectives to revise
• Describe how forces deform objects

Year 8 learning objectives to revise
• Describe the energy stores involved in everyday transfers • Describe the factors that affect the change in temperature of a substance • Compare energy transfer by conduction, convection, and radiation

Pinchpoint focus

Energy stores and conservation of energy.

Tier 2 vocabulary	Tier 3 vocabulary
energy, heating	thermal equilibrium, thermal store

Trigger interest

Show students the image from the Lesson presentation of a flower resting on a piece of aerogel. Ask: *How is this possible?* Lead discussion to the fact that aerogel is an extremely good insulator. It is (in simple terms) a hardened foam; it is 99.8% trapped air. So energy cannot transfer well by conduction (air is a poor conductor) or by convection (because there is no air circulation).

Show a video of someone starting a fire with a magnifying glass. Ask students to explain why the fire starts in terms of energy. (Light carries energy from the Sun. The lens concentrates this energy on a small spot, so that spot gets very hot.)

Recall

Remind students that energy can be stored in different kinds of stores and transferred between stores in different ways.

Give students the short quiz revising energy and thermal conductivity on the Lesson presentation.

Ask students to draw cartoons to show how energy transfers by conduction, convection, and radiation.

List the different renewable energy resources and ask students to brainstorm everything they can remember about each one.

Main activity

Pinchpoint: Energy (20–25 mins)

The pinchpoint question is about a metal spoon (at room temperature) dropped into a pan of hot water. It tests students' understanding of why objects and materials change temperature.

Students answer the pinchpoint question and complete the follow-up activity with the same letter. They then complete the *What I need to remember* activity. Information about each incorrect answer's misconceptions can be found on the teacher sheet. You may want to divide the class into groups, depending on the letter answer chosen, to support each other when completing the follow-up activity.

Ask students to write a story involving as many energy stores as possible.

Allow students to shine a lamp onto a solar panel, then vary the current through the lamp and measure the voltage across the solar panel.

Review and reflect

Students are shown the pinchpoint question again and asked to reflect on whether they think they chose the correct letter and if they would change their answer now. If they think they got it wrong, ask them to think about why they chose that answer.

Reveal the correct answer. Review with students why it is correct and the other options are not. Guidance can be found on the teacher sheet. Students could be asked to pick an incorrect answer and write an explanation to a fictional student who chose it, explaining why it cannot be correct.

Invite students to read their stories. The class can list the energy stores mentioned and give a score. Correct as necessary.

Guide students to explain their results using energy stores. You can introduce the idea of efficiency – was all the energy from the power supply transferred to the solar panel?

Homework

Ask students to complete the knowledge organiser.

Theme: Energy

Sub-theme: Energy stores

Physics NC links

- heating and thermal equilibrium: temperature difference between two objects leading to energy transfer from the hotter to the cooler one, through contact (conduction) or radiation; such transfers tending to reduce the temperature difference: use of insulators
- other processes that involve energy transfer: changing motion, dropping an object, completing an electrical circuit, stretching a spring, metabolism of food, burning fuels
- energy as a quantity that can be quantified and calculated; the total energy has the same value before and after a change
- comparing the starting with the final conditions of a system and describing increases and decreases in the amounts of energy associated with movements, temperatures, changes in positions in a field, in elastic distortions and in chemical compositions
- using physical processes and mechanisms, rather than energy, to explain the intermediate steps that bring about such changes

Working scientifically NC link

- understand that scientific methods and theories develop as earlier explanations are modified to take account of new evidence and ideas, together with the importance of publishing results and peer review

Learning objectives

- Analyse situations in terms of energy stores
- Know how ideas about energy have changed.

Year 7 and 8 learning objectives to be applied

- Describe the energy stores involved in everyday transfers
- Describe the ways of transferring energy between stores
- Use the law of conservation of energy in energy analysis

Applications

- Ideas about energy change over time
- Situations where conservation seems to be violated.

Tier 2 vocabulary	Tier 3 vocabulary
	elastic, energy store, gravitational potential energy (GPE), kinetic

SB in-text question answers

A Chemical **B** The bars on the left are equal to the bars on the right **C** Concrete as you can experience going faster and slower

Lesson plan

Reactivate knowledge

1 Name the energy store that fills when you lift something up.
2 Name the energy store that empties when something slows down.
3 Name the energy store that means the Sun shines.

Answers: 1 Gravity/gravitational potential **2** Kinetic **3** Nuclear

Lesson resources

Activity: *Design a Rube Goldberg machine* (student sheet, support sheet, teacher and technician sheet)

Trigger interest

Ask: *Why do we feel like we have more energy after we have a rest?* (Answer: Whilst resting, our bodies have broken down food to replenish the sugar in our blood.) Facilitate a discussion, and stress that we get energy from our food.

Exposition of main content

Abstract concepts Explain that energy is an abstract concept – we cannot see or touch it. So, it has taken a long time for scientists to define energy. Remind students of the law of conservation of energy: Energy is never created or destroyed. It is only transferred between different stores.

Ask: *What do scientists do to help visualise abstract concepts?* Guide discussion to the use of 'models'. For example, water in jugs can represent energy moving between stores.

Using jugs of water, model the energy changes for:

- a rollercoaster
- a 'popper' toy (you could also demonstrate a popper toy in action)
- a ball falling, bouncing, and travelling back up towards its original height.

Demonstrate the ball bounce and ask: *Why doesn't the ball bounce back up to its original height?*

Model the energy transfers one more time, pouring a little bit of water into a jug labelled 'thermal store' all the way through.

Rube Goldberg machine Show students a video of a Rube Goldberg machine, such as the music video for OK Go's 'This Too Shall Pass'.

Ask students: *Can you identify different energy stores, and the transfers between stores? Can you identify points where energy is added?*

Main activity

Activity: Design a Rube Goldberg machine (20 mins) Students work individually or in pairs to design (on paper) their own Rube Goldberg machine. Ask them to label as many energy stores as they can and add arrows to show energy transferring between stores.

Demonstrate how to draw this for a domino falling; for example, a picture of a domino balanced on its end, and labelled 'gravitational store → kinetic store'.

Review and reflect

Students show their partners their machine designs. If students have attempted the extension, their partners can give them a score for how many different energy stores they included. Ask students to share with the class their favourite energy transfer from their partner's work.

Homework

Students identify energy transfers in their daily lives and in the world around them that include all of the energy stores on the homework slide.

2.3 Apply: Energy transfers

Theme: Energy

Physics NC link

- heating and thermal equilibrium: temperature difference between two objects leading to energy transfer from the hotter to the cooler one, through contact (conduction) or radiation; such transfers tending to reduce the temperature difference: use of insulators

Sub-theme: Energy transfer

Working scientifically NC links

- pay attention to objectivity and concern for accuracy, precision, repeatability and reproducibility
- evaluate risks
- make predictions using scientific knowledge and understanding
- select, plan and carry out the most appropriate types of scientific enquiries to test predictions, including identifying independent, dependent and control variables, where appropriate
- use appropriate techniques, apparatus, and materials during fieldwork and laboratory work, paying attention to health and safety
- make and record observations and measurements using a range of methods for different investigations; and evaluate the reliability of methods and suggest possible improvements
- present observations and data using appropriate methods, including tables and graphs
- interpret observations and data, including identifying patterns and using observations, measurements and data to draw conclusions

Learning objectives

- Analyse situations in terms of energy transfers
- Explain different rates of thermal energy transfer and how to investigate them.

Year 7 and 8 learning objectives to be applied

- Describe how energy is transferred by particles during conduction
- Describe how energy is transferred by particles during convection
- Explain how an insulator can reduce energy transfer
- Give the definition of radiation

Application

- Getting accurate thermal conductivity measurements.

Tier 2 vocabulary	Tier 3 vocabulary
conduction, range	control variable, infrared radiation, thermal conductivity, thermal equilibrium

SB in-text question answers

A The blocks on the right will feel colder **B** They reflect (infrared) radiation from the Sun/do not absorb radiation **C** The temperature of the block will fall, but only slightly

Lesson plan

Reactivate knowledge

1 Name the ways of transferring energy that involve particle motion.
2 Name the type of energy transfer that can happen in a vacuum.
3 Name the other two types of energy transfer.

Answers: 1 Conduction, convection **2** Radiation **3** Forces, electric current

Lesson resources

Practical: *Investigating the effect of temperature difference in heating* (student sheet, support sheet, teacher and technician sheet)

Trigger interest

Elicit students' prior knowledge of thermal insulators to explain how astronauts are protected from the high temperatures of the outside of the space capsule.

Use the Lesson presentation slides to recap the difference between temperature and (heat) energy, thermal equilibrium, and conduction.

Exposition of main content

Pass around pieces of metal and wood. Ask: *Why does the metal feel colder?* Guide discussion towards the fact that metal has higher thermal conductivity than wood (some materials allow energy to transfer through them by conduction quicker than others do).

Remind students that the **thermal conductivity** of a material is how much energy it transfers per second through $1\,m^2$.

Ask: *What else affects the speed of energy transfer from one material to another by conduction?*

Guide discussion towards the fact that the temperature difference between the two materials has an effect.

Main activity

Practical: The effect of temperature difference in heating (30 mins) Ask students to plan and carry out an investigation into the effect of temperature difference on heating.

Steer them towards putting freshly boiled water into test tubes and placing these in beakers filled with water at different temperatures, then measuring the temperature of the water in the test tubes over, say, ten minutes.

Discuss what control variables students think they will need. Remind them of the importance of choosing suitable ranges. How many measurements will they take? How will they get a big enough range of temperature differences? Do they need to take safety precautions?

Students can plot a single graph of all their results, with a line of best fit for each test tube.

Review and reflect

Ask students to use the questions on the Lesson presentation to help them draw conclusions and evaluate their investigation, and to discuss with a partner. Then facilitate a whole-class discussion.

Homework

Students write conclusions and evaluations for their investigations.

2.4 Apply: Energy resources

Theme: Energy

Physics NC links

- fuels and energy resources
- energy as a quantity that can be quantified and calculated; the total energy has the same value before and after a change
- comparing the starting with the final conditions of a system and describing increases and decreases in the amounts of energy associated with movements, temperatures, changes in positions in a field, in elastic distortions and in chemical compositions
- using physical processes and mechanisms, rather than energy, to explain the intermediate steps that bring about such changes

Sub-theme: Energy resources

Working scientifically NC links

- ask questions and develop a line of enquiry based on observations of the real world, alongside prior knowledge and experience
- present reasoned explanations, including explaining data in relation to predictions and hypotheses

Learning objectives

- Describe electricity generation
- Describe the effect of using electrical devices on energy resources.

Year 7 and 8 learning objectives to be applied

- Describe the difference between a renewable and a non-renewable energy source
- Describe how energy is generated with renewable and non-renewable energy sources
- Describe the ways of transferring energy between stores

Application

- Doing an energy analysis of phone charging and use.

Tier 2 vocabulary	Tier 3 vocabulary
energy resource, non-renewable	energy analysis diagram

SB in-text question answers

A For example, toaster, oven, microwave, blender, food processor **B** You can grow trees quickly compared to making coal **C** For example, burning coal

Lesson plan

Reactivate knowledge

1. Name three fossil fuels.
2. Define 'renewable'.
3. Name two things in a thermal power station that spin.

Lesson resources

Activity: *Energy analyses* (information sheet, student sheet, support sheet, teacher and technician sheet)

Answers: 1 Oil, coal, gas **2** Will not run out (within human life spans) **3** Turbine, generator

Trigger interest

Ask: *When you are thinking about the start and end points of an energy analysis for an electrical device, how far 'back' do you go? To the socket? What is behind that?*

Discuss the importance of producing electricity, and its effect on the environment.

Exposition of main content

Talk through the energy analysis diagrams for a phone charging (via a coal power station) and a phone being used on the Lesson presentation slide. Stress how they show the conservation of energy.

Ask: *Can you recall seven renewable energy resources?*

Facilitate discussion until these seven are mentioned: wind, wave, tidal, hydroelectric, geothermal, solar, biomass. Highlight the differences between wave, tidal, and hydroelectric.

Main activity

Activity: Energy analyses (35 mins) Ask students to complete the energy analysis diagrams for the seven renewable energy resources and a variety of household appliances.

Role-play Divide students into science advisors and government ministers. You may decide to do this as a whole-class activity or in smaller groups. The ministers are deciding whether to support spending some of their budget on building solar roads in your country. (You can choose other means of electricity production, instead of solar roads.) Students who are ministers draw up a list of questions they have for their science advisors. Science advisors research the pros and cons of solar roads. Students then role-play the meeting between ministers and advisors.

Review and reflect

Ask students to write a paragraph stating their opinion on whether the ministers should support building more solar roads, and justifying it.

Homework

Students research a renewable energy resource in use (or proposed/under planning) in their areas.

2.5 Apply: Conservation and dissipation

Theme: Energy

Sub-theme: Energy transfer

Physics NC links

- comparing the starting with the final conditions of a system and describing increases and decreases in the amounts of energy associated with movements, temperatures, changes in positions in a field, in elastic distortions and in chemical compositions
- using physical processes and mechanisms, rather than energy, to explain the intermediate steps that bring about such changes

Working scientifically NC link

- interpret observations and data, including identifying patterns and using observations, measurements and data to draw conclusions

Learning objectives

- Use energy conservation and dissipation in energy analyses
- Describe methods to reduce dissipation.

Year 7 and 8 learning objectives to be applied

- Use the law of conservation of energy in energy analysis
- Apply the conservation of energy to simple machines

Application

- Energy analysis in terms of dissipation and making power stations more efficient.

Tier 2 vocabulary	Tier 3 vocabulary
insulation	combined heat and power, dissipation, law of conservation of energy

SB in-text question answers

A It would be higher **B** The table/air
C Any combustible fuels: coal, oil, gas, biofuels

Lesson plan

Reactivate knowledge

1. Write down the law of conservation of energy.
2. Define 'dissipation'.
3. Describe one way to keep a house warmer.

Lesson resources

Practical: *Calculating energy dissipated* (student sheet, support sheet, teacher and technician sheet)

Answers: 1 Energy cannot be created or destroyed, only transferred **2** Transfer to a less useful store **3** Thick curtains/carpets/wall or loft insulation/double glazing

Trigger interest

Ask: *Is the water at the bottom of a waterfall warmer or cooler than the water at the top? What does this tell us about energy?* Guide discussion to the correct answer (water is warmer at the bottom since gravitational energy is transferred to a thermal store).

Exposition of main content

Energy dissipation Ask students to rub their hands together and ask what they notice. They should mention that their hands feel warmer. Ask: *Can you think of other situations where energy from a kinetic store is transferred into the thermal store of the surroundings?* (For example, the temperature increase of an eraser after they have rubbed something out).

Ask: *Can you recall the law of conservation of energy? What is dissipation?* Facilitate discussion, and highlight the point that, when ***any*** object experiences energy transfers, some energy transfers to the thermal stores of the object and its surroundings. This is called energy dissipation. Generally, this energy is wasted.

Use the energy analysis diagrams on the Lesson presentation as examples, then ask: *How can you reduce energy dissipation in these examples?*

You could demonstrate dissipation by bouncing a ball. The ball will not rebound to its original height because some of the energy in its gravitational store is dissipated to the thermal store of its surroundings.

Main activity

Practical: Calculating energy dissipated (30 mins) Before students begin the practical, introduce the formula for gravitational potential energy (GPE). Calculate the initial GPE of the first object to be tested in the practical, as an example.

Then ask students to complete the practical. They drop objects and measure the original and rebound heights. They then calculate the energy dissipated by calculating the difference in gravitational potential energy at the two heights.

Review and reflect

Ask students to imagine that they are children's toy designers. They should discuss in groups what toys they might design that would need materials that:

- hardly dissipate any energy at all
- dissipate some energy, but not all of it
- dissipate lots of energy.

Discuss ideas as a class.

Homework

Students research an example of how energy dissipation is reduced in their daily lives.

2.6 Apply: Work and power

Theme: Energy

Sub-theme: Power and work

Physics NC link

- using physical processes and mechanisms, rather than energy, to explain the intermediate steps that bring about such changes

Working scientifically NC link

- apply mathematical concepts and calculate results

Learning objective

- Analyse situations and do calculations involving work and power.

Year 7 and 8 learning objectives to be applied

- Give the definition of work
- Calculate work done
- Describe the difference between power and energy
- Calculate power and energy

Application

- Combining equations for work and power.

Tier 2 vocabulary	Tier 3 vocabulary
	power, work

SB in-text question answers

A Since the force from the Earth on the Moon is at right angles to the distance moved by the Moon around the Earth **B** 360 J **C** 10 W

Lesson plan

Reactivate knowledge

Lesson resources

Practical: *Calculating power* (student sheet, support sheet, teacher and technician sheet)

1. Write down the equation for work.
2. Write down the equation for power.
3. Write down the units of energy, work, and power.

Answers: 1 Work = force × distance **2** Power = energy/time **3** Energy J, work Nm = J, power W

Trigger interest

Ask: *When you use your muscles to hold a weight in the air, are you doing work?* Guide discussion towards the definition of 'work done' as 'energy transferred to an object by a force'. (So, since holding a weight stationary does not transfer energy to the weight, you are not doing work.)

Exposition of main content

Doing work Ask: *Does the Earth's gravity do work on the Moon?*

Guide discussion towards the answer: No, because if an object doesn't move in the direction of the force, then no work is done. No energy is transferred into or out of the Moon's gravitational and kinetic energy store.

Work and power Ask students to imagine two people, both the same weight, doing press-ups. One person does 30 press-ups in one minute and another person does 30 press-ups in two minutes. They do the same work. So, what is the difference?

Guide discussion to the answer that the difference is how quickly the two people did the work (transferred energy). Define power as the *rate of work done.*

Main activity

Practical: Calculating power (35 mins) Ask students to complete the power calculations.

First, work through the calculation on the Lesson presentation (also in the Student Book) to remind students how to calculate power.

Students lift bean bags one at a time from the floor to the table, as many times as they can in a minute. They then calculate the total work done on their bean bag and their useful power.

Finally, students complete work done and power calculations from the worksheet.

Review and reflect

Students peer-mark their calculations.

Homework

The power students calculated in Practical: *Calculating power* is predominantly the power of their arms. Students follow the instructions on the back of the worksheet to work out the power of their legs, and compare their arms and legs.

2.7 Reach: Efficiency

Theme: Energy

Physics NC links

- comparing amounts of energy transferred (J, kJ, kW hour) (KS3)
- domestic fuel bills, fuel use and costs (KS3)
- fuels and energy resources (KS3)
- calculating energy efficiency for any energy transfers (KS4)
- renewable and non-renewable energy sources used on Earth, changes in how these are used (KS4)

Sub-theme: Energy efficiency

Working scientifically NC link

- explaining everyday and technological applications of science; evaluating associated personal, social, economic, and environmental implications; and making decisions based on the evaluation of evidence and arguments (KS4)

Learning objective

- Describe what is meant by 'efficient' and calculate 'efficiency'.

Year 7 and 8 learning objectives to be built on

- Use the law of conservation of energy in energy analysis

Key Stage 4 link

- Calculations in terms of energy and power.

Tier 2 vocabulary	Tier 3 vocabulary
efficiency	

SB in-text question answers

A 70% **B** 60%
C The law of conservation of energy

Lesson plan

Reactivate knowledge

1. What are the units of energy and power?
2. What is dissipation?
3. What is a non-renewable resource?

Answers: 1 Joules, watts **2** Transferring energy in a way that is not wanted, or to a store that is not wanted **3** One that will run out in a human timescale

Lesson resources

Activity: *Efficiency calculations and extended writing* (student sheet, support sheet, teacher and technician sheet)

Trigger interest

Ask: *Can we make non-renewable energy resources last longer?*

Guide discussion towards the idea that we can, by using more efficient appliances (ones that don't waste as much energy).

Use the diagrams on the Lesson presentation to illustrate how a more efficient appliance achieves the same amount of useful work for less input energy.

Make the link between reducing energy usage from fossil fuels burning and reducing global warming.

Exposition of main content

Energy transfers in a car Ask students to discuss, in pairs, which energy stores are the 'input', 'useful output', and 'wasted output' for a car.

Calculating efficiency Work through example calculations, using the efficiency formula for the two light bulbs.

Main activity

Activity: Efficiency calculations and extended writing (30 mins) Ask students to complete the efficiency calculations.

Ask students to plan (in pairs) and write (individually) an answer to the 6-mark GCSE-style question:

Why do more efficient appliances benefit the environment?

Review and reflect

Students peer- or self-mark their answers.

Homework

Students identify three ways they can increase efficiency in their own homes.

2.8 Reach: Elastic energy and Hooke's Law

Themes: Energy, Forces and motion

Sub-themes: Deformation and Hooke's Law, Energy stores

Physics NC links

- energy as a quantity that can be quantified and calculated; the total energy has the same value before and after a change (KS3)
- using physical processes and mechanisms, rather than energy, to explain the intermediate steps that bring about such changes (KS3)
- energy changes in a system involving heating, doing work using forces, or doing work using an electric current: calculating the stored energies and energy changes involved (KS4)

Working scientifically NC link

- use and derive simple equations and carry out appropriate calculations

Learning objectives

- Describe what elastic energy depends on
- Calculate elastic energy using an equation or a graph.

Year 7 and 8 learning objectives to be built on

- Apply Hooke's law
- Describe the energy stores involved in everyday transfers

Key Stage 4 link

- Calculations using equations and graphs.

Tier 2 vocabulary	Tier 3 vocabulary
area, deform, extension	elastic energy, elastic limit, elastic store, limit of proportionality, spring constant

SB in-text question answers

A A stiffer spring needs a bigger force to produce the same extension, so F/x is bigger **B** The energy doubles **C** Roughly 7.5 N

Lesson plan

Reactivate knowledge

1 What is Hooke's Law?
2 What is the difference between a linear and non-linear material?
3 What is the elastic limit?

Lesson resources

Practical: *Finding the energy stored in a stretched spring* (student sheet, support sheet, teacher and technician sheet)

Answers: 1 The extension is proportional to the force **2** Linear materials obey Hooke's Law, non-linear ones do not **3** Beyond a certain force/extension, the extension is no longer proportional to the force, which is the elastic limit

Trigger interest

Jump like a kangaroo Ask: *What affects how high humans can jump?*

Guide discussion towards the answer: Your mass and how strong your leg muscles are.

Ask: *Why do you think a kangaroo can jump higher than a human, if it keeps jumping?*

Guide discussion towards the answer: Kangaroos have much bigger Achilles tendons, which can store more elastic energy on landing.

Point out that the running blades of Paralympic sprinters work in the same way.

Exposition of main content

Recalling Hooke's Law Ask: *Can you remember what the following terms mean: extension, Hooke's Law, linear material, spring constant?* Guide discussion towards the answers in the Lesson presentation notes. Remind students of the variables in Hooke's Law.

Introduce how elastic energy stored is calculated.

Note that the area is the total force × extension, as the force changes. Use the line for the spring on the Lesson presentation to model finding the energy from the graph. Invite students to explain how finding the energy from the graph explains the energy formula. Model deriving the energy formula from Hooke's Law and the graph, as in the Student Book.

Ask: *Why do you think metals, rubber, and plastic show different patterns of stretching?* Guide discussion towards the fact that these materials have different molecular structures.

Main activity

Practical: Finding the energy stored in a stretched spring (30 mins) Ask students to complete the practical, where they obtain a force–extension graph for loading a spring. They then calculate the energy stored in the spring using the graph. If time permits, students can repeat for a rubber band and/or plastic strip.

Review and reflect

Ask students to compare the scatter of points on their graphs with that on other students' graphs, and suggest sources of random error that have caused it.

Homework

Students research a use of springs in engineering. They should suggest why storing energy in the spring is important for this use.

Introduction to checkpoint intervention

This checkpoint intervention provides suggestions for a lesson to follow up the P3 2 *Energy* checkpoint assessment. Depending on the outcomes of the assessment, these suggestions could move a class towards secure, to target specific learning outcomes, or to consolidate knowledge for students achieving secure outcomes. Use students' outcomes from the checkpoint assessment to address general misconceptions from the content covered. Be prepared to re-cover content as required; three key concepts from the chapter are covered in more detail below.

Checkpoint secure learning outcomes

- Analyse situations in terms of energy stores.
- Know how ideas about energy have changed.
- Analyse situations in terms of energy transfers.
- Explain different rates of thermal energy transfer and how to investigate them.
- Describe electricity generation.
- Describe the effect of using electrical devices on energy resources.
- Use energy conservation and dissipation in energy analyses.
- Describe methods to reduce dissipation.
- Analyse situations and do calculations involving work and power.

Student reflection

Encourage students to reflect on whether there were any checkpoint questions they found difficult or straightforward, as well as their preparation for the checkpoint assessment, including how they revised, the time spent on revision, and what areas they could improve on. For more strategies, see the Metacognition chapter or the Metacognition in Key Stage 3 Science guide on Kerboodle.

Intervention activity: Explain different rates of thermal energy transfer and how to investigate them

Thermal conductivity will be a new concept for most students. It will help if you link this concept with more familiar, concrete situations, for example, metal feeling cold to the touch, even though it is the same temperature as other (non-living) objects in the room. Similarly, students may not have come across the idea that the temperature difference between an object and its surroundings affects how quickly it changes temperature.

Temperature practicals

Allow students to carry out a practical to measure the temperature of water in different test tubes as it cools. The test tubes should be identical, contain the same amount of water, have the tops sealed in the same way, and be sitting in water baths of the same temperature. The water in the test tubes should, however, start at different temperatures (such as 30, 40, 50, 60, and 70 °C). Then measure the temperature of the water in each test tube every minute and plot a graph to show the temperature change. Students should find that all test tubes cool to the same final temperature, but the higher the starting temperature, the steeper the line at any time.

Intervention activity: Use energy conservation and dissipation in energy analyses

Some students find it difficult identifying which energy transfers are useful and which are wasted. Tell students to ask themselves: *What is this appliance trying to do? What energy transfer(s) is that?* When energy is wasted, it is transferred to thermal stores or as sound. Dissipation is when energy is transferred away to the environment and cannot be retrieved; in other words it is 'lost'.

Identifying energy stores

Ask students to practise identifying useful and wasted energy stores for a range of different appliances and actions, such as a television, riding a bike, a toaster. To help students understand dissipation, you can model energy as orange juice. Imagine you are at the seaside and you pour orange juice into your cup, but some spills into the sea and is wasted. The orange juice that went into the sea is still there – it hasn't stopped existing – but you can't get it back. This is like energy dissipating into the environment.

Intervention activity: Analyse situations and do calculations involving work and power

In situations when energy is transferred by a force, many students struggle to use the words 'work done' in place of 'energy transferred'. Many students also struggle to understand the link between exerting less force, doing less work, and having less power, even though they know from real-life experience that these things are related.

Steps and ramps

Take the example of steps and ramps: many public buildings have about eight steps straight up to the front door and a very long, shallow ramp up to it as well, for wheelchair users. Whether you walk up the steps or up the ramp, you do the same amount of work 'against gravity' (that is, in the opposite direction to the force of gravity on you), because you climb the same height overall. But, when you walk up the ramp, you don't have to push yourself upwards as much with each step. The F in the work done formula is smaller, so each step does less work. This means you do work more slowly, that is, your power is less.

P3 Chapter 3: Waves, sound, and light

Introduction to chapter

This chapter allows students to practise explaining physical phenomena by talking about waves. It contains lots of uses of waves, including ultrasound scans, telescopes, and x-rays. Students will also think about the risks associated with waves and how to balance risk and communicate conclusions to the public.

Students will be familiar with sound waves (including ultrasound), (surface) water waves, and visible light; this chapter introduces them to the rest of the EM spectrum.

Students will develop their maths skills by calculating frequency from time period and by calculating the distance to an object in echolocation, but will not use the wave equation.

This chapter includes drawing ray diagrams for the images produced by converging lenses (P3 3.4 *Reflection and refraction*), which is a challenging activity.

Prerequisite knowledge

- The concept of a wave as something that transfers energy from a source
- Sound is a longitudinal wave that travels through solids, liquids, and gases, and can be represented as sinusoidal waves; it has many applications involving reflection
- Waves have amplitude, wavelength, and frequency
- Light is a transverse wave; we represent light as a ray to show its direction
- Humans can only hear/see certain frequencies of sound/light
- Different frequencies of light correspond to us seeing different colours
- White light is made up of all other colours

Core concepts

- Wave properties
- Seeing light

GCSE links

- Calculating wave speed, using frequency and wavelength
- Determining the speed of water waves or sound waves experimentally
- Ultrasound and infrasound
- Seismic waves
- The effects of ionising radiation

Key learning misconceptions

- The amplitude of a wave is the bottom of a wave to the top.
- As the wave travels, the material travels with it.
- Sound is carried by individual molecules.
- Sound and light travel at the same speed.
- Light from a bulb only extends outward a certain distance and then stops.
- Light always passes straight through transparent material without changing direction.
- When objects are seen, light comes out of the eye and travels to the object.
- Light is reflected by shiny surfaces, but not reflected at all from other surfaces.
- All objects that you can see give out their own light.
- All radiation is dangerous and should be blocked.

Teaching preparation

Representing waves

Wavefronts – rays – sine waves – particles It is crucial to recognise that we represent waves in different ways, depending on why we are drawing the diagram. This is dealt with in the pinchpoint activity at the start of the chapter. Making this explicit and clear to students will avoid confusion later on.

Key vocabulary

Wave speed, wavelength, amplitude, frequency, time period These quantities are not conceptually difficult, which is perhaps why students tend to forget them. It is worth including games and quizzes on these terms regularly to improve student recall and avoid confusion when they are used in explanations.

Key maths skill

Using formulae Students may need help rearranging, $f = \frac{1}{T}$, as some may be confused by the '1'. The classic error in echolocation questions is forgetting to double the distance (or halve the time). It is worth reminding students how to convert into standard units (milliseconds to seconds, kilometres to metres), before setting them away on calculations.

Working scientifically

Science communication This chapter allows you to focus on science communication, on how we decide what information to believe, and the potential problems that can arise when people do not understand a piece of science or technology. A suggested example is the vandalism of mobile phone masts during lockdown by people who believed the masts caused and/or spread COVID-19.

3.1 Reactivate: Waves, sound, and light

Theme: Waves

Sub-themes: Light, Sound, Types of wave, Wave properties

Physics NC links

- waves on water as undulations which travel through water with transverse motion; these waves can be reflected, and add or cancel – superposition
- sound produced by vibrations of objects, in loud speakers, detected by their effects on microphone diaphragm and the ear drum; sound waves are longitudinal
- use of ray model to explain imaging in mirrors, the pinhole camera, the refraction of light and action of convex lens in focusing (qualitative); the human eye

Learning objective

- To recall and organise key knowledge about wave properties, sound, and light.

Year 7 learning objectives to revise

- Describe the different types of waves and their features
- Describe what happens when light is reflected
- Describe uses of ultrasound to make images and find distances
- Explain what happens when light is refracted
- Compare a simple camera with the eye
- Describe how primary colours add to make secondary colours

Pinchpoint focus

Methods of representing waves.

Tier 2 vocabulary	Tier 3 vocabulary
compression, frequency	amplitude, longitudinal wave, rarefaction, ray, refraction, transverse wave, wavefront, wavelength

Trigger interest

Show students the image of the bat from the Lesson presentation. Ask: *How do bats use sound waves to catch insects?*

Guide discussion towards the idea of ultrasound; link echolocation to other uses of ultrasound, such as building an image of a baby.

Some blind people have learnt to echolocate. Search online for suitable video clips to show students.

Recall

Remind students that all waves reflect off surfaces. Ask: *Can you think of other things that all waves do? Can you think of anything that all waves have?* Give students the short quiz revising key terms describing waves on the Lesson presentation.

Draw two oscilloscope traces on the board with different amplitudes and time periods. Ask students to suggest differences between the sound waves.

Ask students to order sentences to explain how a ship can work out the distance to a shoal of fish by reflecting an ultrasound pulse off it.

Main activity

Pinchpoint question (20–25 mins)

Waves are important in many areas of our lives. We represent them in different ways, depending on why we are drawing the diagram.

Students answer the pinchpoint question and complete the follow-up activity with the same letter. They then complete the *What I need to remember* activity. Information about each incorrect answer's misconceptions can be found on the teacher sheet. You may want to divide the class into groups, depending on the letter answer chosen, to support each other when completing the follow-up activity.

Ask students to form a line with their hands touching the elbows of the students next to them to form a wave front. Draw a line on the ground at an angle to the line of the students. This is the boundary between air (where the students are) and water. Students walk in time with each other while they are in the 'air', then shuffle on tiptoe once they are in the 'water'. This should cause the wavefront to bend where it enters the water and so change direction from the air to the water, modelling refraction. Students can then describe this model for refraction.

Ask students to stand far from a wall, clap, and work out the speed of sound in air from the time between the clap and its echo being heard.

Review and reflect

Students are shown the pinchpoint question again and asked to reflect on whether they think they chose the correct letter and if they would change their answer now. If they think they got it wrong, ask them to think about why they chose that answer.

Reveal the correct answer. Review with students why it is correct and the other options are not. Guidance can be found on the teacher sheet. Students could be asked to pick an incorrect answer and write an explanation to a fictional student who chose it, explaining why it cannot be correct.

Students can evaluate the model for refraction they have just acted out.

Tell students the correct speed of sound in air (343 m/s for standard temperature and pressure). Students' values from the clapping experiment will obviously be different. Ask them why.

Homework

Ask students to complete the knowledge organiser.

3.2 Apply: Wave properties

Theme: Waves

Sub-themes: Types of wave, Wave properties

Physics NC links

- amplitude, wavelength, frequency, relating velocity to frequency and wavelength (KS4)
- transverse and longitudinal waves (KS4)

Learning objective

- Compare properties of waves, including those of seismic waves.

Year 7 and 8 learning objectives to be applied

- Describe the different types of waves and their features
- Describe how a sound is produced and travels

Application

- Finding the time period of a wave.

Tier 2 vocabulary	Tier 3 vocabulary
frequency, snapshot	amplitude, longitudinal, P-wave, period, S-wave, seismic, time trace, transverse, wavelength

SB in-text question answers

A 4 s **B** A transverse wave is made by moving your hand at right angles; a longitudinal wave is made by moving your hand in and out. **C** $\frac{1}{3}$ s

Lesson plan

Reactivate knowledge

1 Is light transverse or longitudinal?

2 Is sound transverse or longitudinal?

3 What are the areas in a sound wave called where the particles are close together?

Answers: 1 Transverse **2** Longitudinal **3** Compressions

Lesson resources

Activity: *Measuring time periods* (student sheet, support sheet, teacher and technician sheet)

Trigger interest

Show the image of the surface wave after a stone has been dropped in water in the Lesson presentation. Ask students: *How would you measure the wavelength of this surface water wave?*

Guide discussion to the idea of measuring between two peaks or between two troughs. Explain that you'd need a wave ruler in the photograph to be able to do this.

Show the diagrams of longitudinal and transverse waves. Demonstrate both types on a slinky spring. Ask: *What is the similarity between peaks and troughs, and compressions and rarefactions?*

Exposition of main content

Frequency and time period Ask students: *Can you recall what 'frequency' is?*

Use a slinky spring to demonstrate a transverse wave with a frequency of 1 Hz. Ask: *How does frequency relate to the time period?* Repeat for a 2 Hz wave. Show the formulae on the next slide of the Lesson presentation.

Sound waves and oscilloscopes Stick a piece of coloured tape to a point on the slinky spring, then send a longitudinal wave along the spring. Ask: *How does the coloured tape move?* Remind students that sound waves are like the longitudinal wave on the spring, except with air particles moving.

Show the diagrams on the Lesson presentation and note that, when the air particles hit a microphone, the changing pressure is turned into a changing electrical signal with the same frequency. (You could demonstrate with an oscilloscope and microphone if possible.)

Ask students: *Why does the oscilloscope trace look like a transverse wave?* Make a transverse wave with the slinky spring with the coloured tape. Explain that the oscilloscope trace is like a distance–time graph for the point (tape).

Finding the time period from a trace Use the example on the Lesson presentation to model finding the time period from a trace.

Seismic waves Introduce seismic waves and the differences between P-waves and S-waves. Note that an oscilloscope can be used to find the time period of seismic waves.

Main activity

Activity: Measuring time periods (20 mins) Students use microphones attached to oscilloscopes to measure the time period of a note made by a tuning fork.

Review and reflect

Ask students to discuss the questions on the Lesson presentation, then write the answers in their books. They can use the model answer to self-assess.

Homework

Students compare different pairs of waves using Venn diagrams.

3.3 Apply: Sound and its applications

Theme: Waves

Sub-theme: Sound, Uses of waves

Physics NC links

- frequencies of sound waves, measured in hertz (Hz); echoes, reflection and absorption of sound
- sound produced by vibrations of objects, in loud speakers, detected by their effects on microphone diaphragm and the ear drum; sound waves are longitudinal

Learning objective

- Describe how waves are used to investigate structures that cannot be observed.

Year 7 and 8 learning objectives to be applied

- Describe what happens when water waves hit a barrier
- Describe uses of ultrasound to make images and find distances
- Compare the layers of the Earth

Application

- Using waves to discover internal structure of metals, the human body, and the Earth.

Tier 2 vocabulary	Tier 3 vocabulary
	mechanical wave, seismograph, seismometer, transducer, ultrasound

SB in-text question answers

A The pulse would be further to the right on the screen **B** C **C** P-waves arrive after a shorter time

Lesson plan

Reactivate knowledge

1 Give the lowest frequency of ultrasound.

2 What is the difference between reflection and refraction?

3 Put these states in order of increasing speed of sound travel: liquid, gas, solid.

Answers: 1 20 000 Hz, or 20 kHz **2** Reflection is when a wave bounces off a surface, refraction is when it changes direction moving from one medium to another where the speed is slower **3** Gas, liquid, solid

Lesson resources

Activity: *Calculations and letter writing* (student sheet, support sheet, teacher and technician sheet)

Trigger interest

Ask students: *When you drive over a suspension bridge you will see lots of metal cables. How could you see inside a cable to check there are no cracks?* Guide discussion towards the idea of a scan, and the idea of waves reflecting off something and measuring how much reflects back to a scanner.

Exposition of main content

Ultrasound Ask students: *Do you know how an ultrasound scan works?* Discuss, then describe it for them (see Lesson presentation notes). Then ask: *How would this help look for cracks in a metal cable?* Again, discuss then provide the answer. Show the image on the Lesson presentation and explain that ultrasound can also be used to look inside people.

Calculating distance Use the image and values on the Lesson presentation to demonstrate using the speed formula to find the distance.

Main activity

Activity: Calculations and letter writing (30 mins) In part 1, students complete calculations involving distance and time taken for a sound wave to be reflected.

Clarify the meanings of the linking words on the slide of the Lesson presentation or any you have on posters in your room or in exercise books. Ask students to make sentences about cats and dogs using each word.

In part 2, students plan and write a letter to explain what ultrasound is, how an ultrasound scan works, and why it isn't dangerous for the patient. Encourage students to use linking words as appropriate.

Review and reflect

Students read their partners' letters and feed back sentences that include correct use of scientific ideas and linking words.

Homework

Ask students to write a short magazine article explaining what seismic waves are and how scientists worked out the structure of the Earth by detecting them. They should try to use linking words to show the relationships between their ideas.

3.4 Apply: Reflection and refraction

Theme: Waves

Sub-theme: Light, Wave properties

Physics NC links

- the transmission of light through materials: absorption, diffuse scattering and specular reflection at a surface
- use of ray model to explain imaging in mirrors, the pinhole camera, the refraction of light and action of convex lens in focusing (qualitative); the human eye

Learning objective

- Use ray diagrams to explain illusions caused by reflection and refraction.

Year 7 and 8 learning objectives to be applied

- Use the law of reflection to explain how images are formed in the plane mirror
- Describe how a convex lens affects light

Application

- The formation of images by lenses.

Tier 2 vocabulary	Tier 3 vocabulary
magnify, lens, ray	convex, focus, *image*, ray diagram

SB in-text question answers

A You would have to be looking in the mirror to see the image **B** Twice

Lesson plan

Reactivate knowledge

1 What is the law of reflection?

2 What is a real image and a virtual image?

3 What happens when light goes from air into water?

Lesson resources

Activity: *Investigating images formed by convex lenses* (student sheet, support sheet, teacher and technician sheet)

Answers: 1 The angle of incidence is equal to the angle of reflection **2** A real image can be put on a screen, a virtual image can only be seen by looking at a shiny surface/lens **3** It is refracted/changes direction/slows down

Trigger interest

Ask students: *When you look at a shop window you see your reflection, and at the same time you can see what is behind the glass. How can we see both?* Use the diagram on the Lesson presentation to guide discussion towards the answer (see Lesson presentation notes).

Ask students: *Does the light passing through the glass travel in a straight line?* Use the diagrams to guide discussion towards a recap of reflection and refraction.

Exposition of main content

Ask students: *When we look through a convex lens, the image that we see is distorted. Why?* Guide discussion towards the correct answer.

Demonstrate forming a real image of something outside the window. Ask: *What do you notice about the image?* (Answer: the image is smaller, upside down, and real.) Recap what 'real' means.

Demonstrate using a convex lens as a magnifying glass. Recap that virtual images cannot appear on a sheet of paper.

(Note: you could hand out convex lenses so students can form real images and use the lens as a magnifying glass.)

Use the ray diagram on the Lesson presentation to show how convex lenses focus parallel rays to a point. Emphasise the simplification of drawing the light ray bending once at the centre line of the lens.

Model using a ray diagram to explain how both real and virtual images can be formed by a convex lens; demonstrate both.

Main activity

Activity: Investigating images formed by convex lenses (30 mins) Students draw ray diagrams to predict what image will be formed when they hold a convex lens away from a lit bulb.

Review and reflect

Ask students to compare their diagrams with the correct ones shown on the Lesson presentation. Then ask them to write a conclusion for their investigation. They can use the template on the Lesson presentation to help.

Homework

Ask students to write a short paragraph explaining how a magnifying glass works and draw a ray diagram to illustrate their explanation.

3.5 Apply: Applications of reflection and refraction

Theme: Waves

Sub-themes: Light, Uses of waves, Wave properties

Physics NC links

- the transmission of light through materials: absorption, diffuse scattering and specular reflection at a surface
- use of ray model to explain imaging in mirrors, the pinhole camera, the refraction of light and action of convex lens in focusing (qualitative); the human eye

Learning objective

- Use reflection and refraction to explain how telescopes and curved mirrors work.

Year 7 and 8 learning objectives to be applied

- Use the law of reflection to explain how images are formed in a plane mirror
- Explain what happens when light is refracted
- Compare a simple camera with the eye

Application

- The formation of images by refracting telescopes, reflecting telescopes, and curved mirrors.

Tier 2 vocabulary	Tier 3 vocabulary
objective	concave, convex, eyepiece, reflecting telescope, refracting telescope

SB in-text question answers

A 30–50 cm **B** Real **C** Behind the mirror

Lesson plan

Reactivate knowledge

1 What is the focus of a lens?
2 Where do you measure angles from and to on a ray diagram?
3 What is dispersion?

Lesson resources

Practical: *Making a refracting telescope* (student sheet, support sheet, teacher and technician sheet)

Answers: **1** The point where the rays cross **2** A line (the normal) at 90 degrees to the surface **3** The separating of white light into colours when it goes through a prism.

Remind students that ‘focus’ means ‘focal point’

Trigger interest

Ask students: *Galileo's observations of Jupiter changed our view of the Solar System. Telescopes are now looking for evidence of alien life. How do they work?* Guide discussion to the idea that telescopes make magnified images either through reflection or refraction.

Exposition of main content

Refracting telescopes Ask students*: How is a refracting telescope made?* Use Figure 1 in the Student Book to show that it is made from two convex lenses. Note that the earliest telescopes were built this way.

Reflecting telescopes and parabolic mirrors Ask students: *How does a reflecting telescope work?* Use Figure 3 in the Student Book to explain that a reflecting telescope focuses light using a parabolic, concave mirror. If possible, allow students to look at and feel some parabolic, concave mirrors. Show students Figure 4 in the Student Book of how a parabolic mirror forms a real image.

Explain that the James Webb Space Telescope, launched in 2021, is a reflecting telescope.

(You could show an online video from BBC's Bang Goes the Theory.)

Ask students: *Why do you think reflecting telescopes are preferred over refracting telescopes?* Guide discussion towards the correct answer (see Lesson presentation notes).

Ask students: *Can you think of anywhere that you have seen convex or concave mirrors in everyday life?* Discuss answers.

Main activity

Practical: Refracting telescopes (30 mins) In part 1, students make a refracting telescope.

In part 2, students design a product or tool that uses a parabolic mirror. They should make a labelled sketch of their design, and explain how it works and why their design is a good idea.

Review and reflect

Ask students to present their products to their classmates. Encourage their classmates to ask questions.

Homework

Ask students to find a picture taken by the James Webb Space Telescope and write a few sentences explaining what it shows.

3.6 Apply: Light and colour

Theme: Waves

Physics NC links

- light transferring energy from source to absorber leading to chemical and electrical effects; photo-sensitive material in the retina and in cameras
- colours and the different frequencies of light, white light and prisms (qualitative only); differential colour effects in absorption and diffuse reflection

Sub-themes: Light

Working scientifically NC link

- the ways in which scientific methods and theories develop over time (KS4)

Learning objective

- Describe how ideas about light have changed, and explain the link between the colours of the spectrum, primary, and secondary colours.

Year 7 and 8 learning objectives to be applied

- Describe how the eye works
- Describe and explain what happens when light passes through a prism
- Describe how primary colours add to make secondary colours

Application

- Changing ideas about colour and how it is perceived.

Tier 2 vocabulary	Tier 3 vocabulary
pixel, primary colour, secondary colour, spectrum	spectral frequency

SB in-text question answers

A e.g., 'Richard Of York Gave Battle In Vain' **B** He used a prism and the blue light beam did not split into other colours **C** Absorbs green and allows blue and red light through

Lesson plan

Reactivate knowledge

1 What do you use to produce a white light spectrum?

2 What are the primary colours of light?

3 How do filters work?

Answers: 1 A prism **2** Red, green, blue **3** They allow certain frequencies/colours of light through and absorb the rest

Lesson resources

Activity: *Investigating colour shadows* (student sheet, support sheet, teacher and technician sheet)

Trigger interest

Ask students: *Why are there seven colours in the spectrum, but only three primary colours of light?* Guide discussion to the fact that our eyes can only detect three colours of light.

Show the diagrams on the Lesson presentation. Ask if they know what the images show. Discuss how a 'red' cone cell, for example, detects red light strongly, and weakly detects the colours either side of red light in the spectrum.

Exposition of main content

Ask students: *So how do we see other colours?*

Use equipment or the online PhET colour vision simulation to demonstrate how mixing primary colours allows us to see:

- the secondary colours
- all the colours that we see.

Emphasise that there are two ways we can see yellow: if yellow light enters our eyes, or if red and green light enters our eyes.

Discuss the discoveries of Ibn al-Haytham and Newton (see Lesson presentation notes). Use the next slide to discuss how, more recently, scientists have explained why we see magenta.

Ask: *Why do leaves look green?,* Guide discussion to the correct answer. Then ask: *What happens if you mix red, green, and blue paint? Why?* Guide discussion to the correct answer, emphasising that mixing paint is not the same as mixing light. (This is why the primary colours in art are not the same as the primary colours for our eyes: if you mix green and purple light, you would see white, because all the colours would be entering your eye.)

Main activity

Activity: Investigating colour shadows (30 mins)
Students investigate how coloured shadows are formed and use their understanding to create specific colour shadows.

Review and reflect

Ask students to view each other's pictures from the far side of the classroom and explain how they achieved different colours.

Homework

Ask students to write a magazine article explaining how a television screen displays a picture that we can see.

3.7 Reach: The electromagnetic spectrum

Theme: Waves

Sub-theme: Electromagnetic waves

Physics NC links

- electromagnetic waves, velocity in vacuum; waves transferring energy; wavelengths and frequencies from radio to gamma rays (KS4)
- uses in the radio, microwave, infrared, visible, ultraviolet, X-ray, and gamma-ray regions, hazardous effects on bodily tissues (KS4)

Working scientifically NC links

- explaining everyday and technological applications of science; evaluating associated personal, social, economic, and environmental implications; and making decisions based on the evaluation of evidence and arguments (KS4)
- evaluating risks both in practical science and the wider societal context, including perception of risk (KS4)
- recognising the importance of peer review of results and of communication of results to a range of audiences (KS4)

Learning objective

- Name the waves of the electromagnetic spectrum, and describe how and why some of them are dangerous.

Year 7 and 8 learning objectives to be built on

- Describe the different types of waves and their features
- Describe how light is emitted, travels, and is detected or absorbed
- Describe what happens when light interacts with matter

Key Stage 4 link

- Electromagnetic waves, ionisation

Tier 2 vocabulary	Tier 3 vocabulary
mutation	electromagnetic spectrum, ionisation, radiation

SB in-text question answers

A Radio, microwave, infrared, visible, ultraviolet, X-rays, gamma **B** The atmosphere allows visible light through, but not gamma rays

Lesson plan

Reactivate knowledge

1 What are the colours of the white light spectrum?
2 What is frequency, and its unit?
3 What is wavelength, and its unit?

Answers: 1 Red, orange, yellow, green, blue, indigo, and violet **2** Number of waves per second, Hertz **3** Distance from one point on a wave to the same point on the next wave in metres

Lesson resources

Activity: *Are mobile phone masts dangerous?* (student sheet, support sheet, teacher and technician sheet), Homework: *Comparing the discoveries of infrared and ultraviolet* (homework sheet)

Trigger interest

Show students the image on the Lesson presentation; it shows a model human torso used by NASA to test the effects of radiation from the Sun on humans on the ISS. Ask: *Do the astronauts on the International Space Station need to wear sunscreen?* Discuss with the class, but don't provide the answer as they will revisit this later. Then ask: *What are the different types of light and what are their dangers?* and briefly discuss before moving onto the Exposition section.

Exposition of main content

Use the Lesson presentation slide to introduce light as an EM wave. EM waves are transverse and do not need a medium to travel through, which means they can travel through a vacuum.

Different frequencies, different properties Explain that the frequency of the wave determines what its properties are.

Dangers of EM waves Introduce each EM wave in turn (spread across two Lesson presentation screens). Link each type to something students can relate to, such as mobile phones and microwave ovens, and explain its dangers. Ask the Trigger interest question about astronauts needing sunscreen again. Using the graph on the Lesson presentation, guide discussion to the correct answer (see Lesson presentation notes).

Main activity

Activity: Are mobile phone masts dangerous? (30 mins)
Ask students to complete the activity – a reading exercise on the controversy around mobile phone masts.
Remind students that mobile phone masts were vandalised during the beginning of the pandemic, because conspiracy theorists thought they were being used to spread COVID-19. Discuss how we decide what to believe, when it comes to science-related questions, and what causes people to believe conspiracy theories.

Review and reflect

Ask students to write three questions (with answers) that all start with:
'Name two electromagnetic waves that …'
They can then swap questions with a partner and mark each other's answers.

Homework

Ask students to complete Homework: *Comparing the discoveries of IR and UV* – a reading exercise about Herschel's and Ritter's discoveries of IR and UV, with questions on the similarities and differences between the two discoveries.

3.8 Reach: Applications of electromagnetic waves

Theme: Waves

Sub-theme: Electromagnetic waves, Uses of waves

Physics NC links

- electromagnetic waves, velocity in vacuum; waves transferring energy; wavelengths and frequencies from radio to gamma rays (KS4)
- uses in the radio, microwave, infrared, visible, ultraviolet, X-ray, and gamma-ray regions, hazardous effects on bodily tissues (KS4)

Learning objective

- Describe some uses of the waves of the electromagnetic spectrum, and reasons for those uses.

Year 7 and 8 learning objectives to be built on

- Describe how light is emitted, travels, and is detected or absorbed
- Describe what happens when light interacts with matter

Key Stage 4 link

- Carrier waves vs signal waves, interaction of waves with matter.

Tier 2 vocabulary	Tier 3 vocabulary
aerial, radioactive, signal	carrier wave, radiotherapy radioactive, radiotherapy, signal

SB in-text question answers

A They do not have enough energy to knock electrons out of atoms. **B** The bigger the current the more intense the microwaves **C** It would damage cells/DNA, which could cause cancer

Lesson plan

Reactivate knowledge

1 How do you use echoes to measure distance?

2 What are the waves of the electromagnetic spectrum?

3 Which waves have a heating effect?

Answers: 1 Send a pulse of sound towards an object, measure the time of the echo, calculate the distance using speed × half the time of the echo **2** radio, microwave, infrared, visible, ultraviolet, X-rays, gamma **3** radio, microwaves, infrared, visible

Lesson resources

Activity: *Electromagnetic waves* (student sheet, support sheet, teacher and technician sheet)

Trigger interest

Ask students: *Different waves of the electromagnetic spectrum have different properties. How can we use them?* Discuss their ideas.

Exposition of main content

Use the Lesson presentation slides to introduce and discuss the uses of EM waves.

Waves carrying information Ask: *Do you know what the following parts of the electromagnetic spectrum are used for, in communication?* Discuss the images – a radio, a mobile phone, and a television remote – and how they use EM waves. Then talk through the following two slides, which give more information about mobile phone signals and transmitters.

Waves used to cook food Ask: *Do you know what the following parts of the electromagnetic spectrum are used for, in cooking?* Discuss the images – microwaves vibrate water molecules in food, and hot objects (such as a toaster wire) emit IR waves.

Note: if there is time you could find the wavelength of the microwaves produced by a microwave oven: remove the turntable and microwave cheese slices laid over kitchen towel for a few seconds. The spots where the cheese melts most are the peaks, so the distance between two melted spots is half a wavelength.

Waves used in medicine Ask: *Do you know what the following parts of the electromagnetic spectrum are used for, in medicine?* Discuss the images – an X-ray and a brain scan – and how they use EM waves. Explain that ionisation is the process of electrons being removed from atoms and that ionisation can cause mutations and lead to cancer when it occurs in living cells. Then ask: *How can we use X-rays and gamma rays to scan people when they are both ionising? Why won't this give you cancer?* Discuss the idea of low doses, and explain that medical professionals have to protect themselves from over-exposure. Finally, describe radiotherapy and explain that the source of the gamma rays is rotated around the patient, so that the healthy cells are damaged less.

Note: UV light is not included on the Lesson presentation, but you can note that it is used for security markings. If you have UV torches, you can pass these around with bank notes and messages written with UV pens.

Main activity

Activity: Electromagnetic waves (25 mins) In part 1, students plan (in pairs), discuss (as a class), and write (individually) an answer to this exam-style question: *Compare the uses and dangers of X-rays and ultrasound waves.*

In part 2, students complete a 'treasure hunt' by matching waves to their descriptions, copying down the clues as they go.

Review and reflect

Ask students to peer assess each other's answers using the mark scheme on the Lesson presentation. They should write comments to help their partner improve their answer.

Homework

Ask students to improve their answer based on their partner's feedback.

P3 Chapter 3: Waves, sound, and light

Introduction to checkpoint intervention

This checkpoint intervention provides suggestions for a lesson to follow up the P3 3 *Waves, sound, and light* checkpoint assessment. Depending on the outcomes of the assessment, these suggestions could move a class towards secure, to target specific learning outcomes, or to consolidate knowledge for students achieving secure outcomes. Use students' outcomes from the checkpoint assessment to address general misconceptions from the content covered. Be prepared to re-cover content as required; three key concepts from the chapter are covered in more detail below.

Checkpoint secure learning outcomes

- Compare properties of waves, including those of seismic waves.
- Describe how waves are used to investigate structures that cannot be observed.
- Use ray diagrams to explain illusions caused by reflection and refraction.
- Use reflection and refraction to explain how telescopes and curved mirrors work.
- Describe how ideas about light have changed.
- Explain the link between the colours of the spectrum, primary, and secondary colours.

Student reflection

Encourage students to reflect on whether there were any checkpoint questions they found difficult or straightforward, as well as their preparation for the checkpoint assessment, including how they revised, the time spent on revision, and what areas they could improve on. For more strategies, see the Metacognition chapter or the Metacognition in Key Stage 3 Science guide on Kerboodle.

Intervention activity: Describe how waves are used to investigate structures that cannot be observed

Students can usually understand the idea of ultrasound waves reflecting off, for example, a crack in a pipe. However, they often struggle to interpret the trace showing the detection of the reflected waves. It is important that students realise that, where there is no crack, the reflected wave has reflected off the *other side of the pipe* and it is by comparison with this trace that you identify unexpected earlier reflected waves.

Scanning a pipe

Simulate scanning a pipe that you have drawn, with a crack or cracks inside. Tell the students how wide the pipe is and the speed of the ultrasound in the pipe. Ask students to 'fire ultrasound waves' from different positions behind a line or bench 'into the pipe'. Tell them how many reflected waves they detect and after how long. Students can test different spots and build up a picture of the crack.

Intervention activity: Use ray diagrams to explain illusions caused by reflection and refraction

Some students get confused by the different ways we can represent waves. When we are only interested in the *direction* the wave is travelling, we represent waves as rays. Only light sources emit light. We see everything else because light reflects off it and into our eyes.

Optical illusions

Demonstrate and draw diagrams for various optical illusions, such as the coin disappearing in the bottom of a beaker of water. Link this to the problem of spear-fishing, and ask students to write an explanation of why the person fishing needs to aim further away than the fish appears to be.

Intervention activity: Explain the link between the colours of the spectrum, primary, and secondary colours

Some students find it counterintuitive that our eyes only detect three colours of light, even though we see a continuous spectrum of colours. Use the online pHet simulator to demonstrate the continuous nature of the spectrum we see, by varying the amounts of red, green, and blue light from the torches.

Red, green, and blue

Ask students to draw their own 'red, green, blue' pixel picture, and then view it from the other side of the classroom. For no red, colour the red pixel black. For some red, colour the red pixel dark red, and for maximum red, colour the pixel red.

P3 Chapter 4: Electricity and magnetism

Introduction to chapter

In this chapter, students revisit ideas from Year 8 with more detail. Applications of phenomena, such as static electricity, link abstract concepts to observable phenomena. Students will cover new uses of series and parallel circuits and explore why some materials are magnetic and others aren't. Heavy use is made of models throughout.

The chapter then introduces some big new ideas: inducing p.d. and the generator. Students will know that you can induce a magnetic field around a wire by passing current through it, but not that you can induce a p.d. across a wire or a current through a loop by moving it past a magnet.

Prerequisite knowledge

- Charges can be positive or negative; like charges repel and opposite charges attract
- Electrons have negative charge
- If you rub two things together, you can rub electrons off one onto the other, making both objects charged
- Current is the charge per second passing a point; potential difference depends on the energy given to a component by the charges passing through it
- Resistance is a measure of how difficult it is for a p.d. to push a current through a component; $V = IR$
- Simple series and parallel circuits
- What a 'field' is, and what shape the magnetic field is around a current-carrying wire and an electromagnet
- Magnetic poles can be north or south; like poles repel and opposite poles attract
- The main parts of a power station

Core concepts

- Forces at a distance
- Static electricity and charge
- Circuits and current

GCSE links

- Uses and dangers of static electricity
- Drawing electric fields
- Magnetic fields around permanent and induced magnets
- Determining the direction of the Earth's magnetic field, using a compass
- Magnetic effects of currents and how solenoids enhance the effect
- How motors and generators work
- How transformers are used in the National Grid and the reasons for their use

Key learning misconceptions

- Objects become positively charged because they have gained protons/electrons have been destroyed.
- All the electrons in an electrical circuit are initially contained in the battery or other source of the electricity.
- Current flows from a battery to a light bulb, but not from the light bulb to the battery.
- Current flows out of both terminals of a battery or power-pack (the 'clashing' current model).
- Current is used up in a circuit.
- The magnetic pole of the Earth in the Northern Hemisphere is a north pole, and the pole in the Southern Hemisphere is a south pole.
- The electrostatic force between two charged objects is not affected by the distance between them.
- Electricity is produced in the wall socket.
- Magnetic field lines are really there – a magnetic field really is a pattern of lines.

Teaching preparation

Teaching strategy

Charge Static charge is introduced first, so that students understand and accept that charge is a real quantity; there are lots of familiar, observable examples of static charge and the Van der Graaf generator is especially memorable. Once students have the concept of charged particles, it is then possible to talk about current (the amount of charge passing a point every second) and p.d. (the energy per unit charge given to a component by each charged particle).

Key vocabulary

Induce Students have met this word before, but it is nonetheless a difficult word for them to begin to use. Give them sentences with 'induce' in (for example, when the current flows in the wire, a magnetic field is 'induced' around the wire) and ask what they think the word means. The closest synonym is probably 'cause'.

Key vocabulary

Modelling This chapter makes regular use of models to support understanding of abstract concepts such as p.d. and magnetism. It is important to stress that models do not describe what is actually happening; rather, they help us to think about what is happening by comparing it to something we can visualise. This is especially important when it comes to the models for circuits in P3 4.3 *Current, p.d., and resistance*; no model is perfect and students will be able to identify questions each model cannot explain.

Hands-on demonstration

The generator The concept of generating a continuous current by rearranging wires and magnets 'just so' and then spinning something is quite difficult for students to grasp. Having some sort of model or large demonstration set would be very beneficial.

4.1 Reactivate: Electricity and magnetism

Themes: Electricity and magnetism

Sub-themes: Charge, Circuits, Currents, resistance, and potential difference

Physics NC links

- electric current, measured in amperes, in circuits, series and parallel circuits, currents add where branches meet and current as flow of charge
- potential difference, measured in volts, battery and bulb ratings; resistance, measured in ohms, as the ratio of potential difference (p.d.) to current.

Learning objective

- To recall and organise key knowledge about electricity and magnetism.

Year 8 learning objectives to revise

- Describe how charged objects interact
- Give the definition of potential difference in terms of energy and force
- Calculate resistance using simple values of potential difference and current
- Describe the difference between series and parallel circuits
- Give the definition of a magnetic field
- Describe how to make an electromagnet
- Describe some uses of electromagnets

Pinchpoint focus

Current and potential difference in circuits.

Tier 2 vocabulary	Tier 3 vocabulary
cell, charge	ammeter, *current*, parallel circuit, potential difference, series circuit, voltmeter

Trigger interest

Show students the hairdryer image from the Lesson presentation. Talk through the bullet points on the Lesson presentation, then ask students to draw the circuit inside the hairdryer.

Recap the meanings of series and parallel.

Draw a series circuit with a cell and a bulb in it. Ask: *What will happen if you add more bulbs in series and in parallel?*

Set up two circuits, one with a cell and two bulbs in series, the other with a cell and two bulbs in parallel. Ask why the bulbs are dimmer when the bulbs are in series and brighter when the bulbs are in parallel.

In series, the total resistance is higher. In parallel, the total resistance is lower, since there is more 'room' for the electrons to flow through.]

Recall

Show students the diagram from the Lesson presentation of a simple series circuit with a cell and a bulb in it. Ask how we measure the current passing through the bulb and what current is, then how we measure the p.d. across the bulb and what p.d. is.

Ask: *What will happen to the current and p.d. if you add more cells/more bulbs?*

Draw a solenoid on the board. Ask students what the magnetic field around the solenoid looks like and how you can make it stronger. (Like a bar magnet: by adding more turns, increasing the current, or adding an iron core.)

Main activity

Pinchpoint: Electricity and magnetism (20–25 mins)

The pinchpoint question is about understanding how current and p.d. behave in series and parallel circuits, and how that allows us to design and improve new appliances.

Students answer the pinchpoint question and complete the follow-up activity with the same letter. They then complete the *What I need to remember* activity. Information about each incorrect answer's misconceptions can be found on the teacher sheet. You may want to divide the class into groups, depending on the letter answer chosen, to support each other when completing the follow-up activity.

Students can investigate the strength of an electromagnet while varying the current in the wire or the number of turns. Strength of magnetic field can be measured at a fixed distance from the solenoid using a magnetometer or approximately by how many paper clips the solenoid can pick up.

Students can use a compass to draw the magnetic field lines around a bar magnet, current-carrying wire, or solenoid.

Review and reflect

Students are shown the pinchpoint question again and asked to reflect on whether they think they chose the correct letter and if they would change their answer now. If they think they got it wrong, ask them to think about why they chose that answer.

Reveal the correct answer. Review with students why it is correct and the other options are not. Guidance can be found on the teacher sheet. Students could be asked to pick an incorrect answer and write an explanation to a fictional student who chose it, explaining why it cannot be correct.

Ask students to compare the methods for measuring magnetic field strength. What are the advantages and disadvantages of each?

Ask what the spacing of lines means. Show students how the Earth's magnetic field deflects harmful charged particles from the Sun, in the process causing the auroras.

Homework

Ask students to complete the knowledge organiser.

Theme: Electricity and magnetism

Physics NC links

- separation of positive or negative charges when objects are rubbed together: transfer of electrons, forces between charged objects
- the idea of electric field, forces acting across the space between objects not in contact

Sub-theme: Charge

Working scientifically NC link

- explaining everyday and technological applications of science; evaluating associated personal, social, economic, and environmental implications; and making decisions based on the evaluation of evidence and arguments (KS4)

Learning objectives

- Explain why sparks happen
- Describe how static electricity can cause problems and how it can be useful.

Year 7 and 8 learning objectives to be applied

- Describe how objects can become charged
- Give the definition of potential difference in terms of energy and force

Application

- Sparks and earthing.

Tier 2 vocabulary	Tier 3 vocabulary
earthing, lightning, spark	ion

SB in-text question answers

A Positive **B** They are charged particles
C Metal, e.g., copper

Lesson plan

Reactivate knowledge

1. What is the charge on a proton and an electron?
2. What do 'like' charges and 'unlike' charges do?
3. Which particles are transferred when you rub a balloon on your clothes?

Answers: 1 Proton – positive, electron – negative **2** Like charges repel, unlike charges attract **3** Electrons

Lesson resources

Activity: *Uses of static electricity* (student sheet, support sheet, teacher and technician sheet)

Trigger interest

Ask: *Have you ever been shocked by a door handle at home, or when reaching for a car door? Can these types of shock be dangerous? Can static electricity be useful?*

Discuss and note students' ideas, but don't explain too much as you will revisit this later.

Exposition of main content

Van der Graaf generator Demonstrate charging the Van der Graaf generator (as outlined in the teacher and technician sheet). Ask: *Can you explain why the dome becomes charged?* Recap like charges repelling, by making a student's hair stand up. Ask: *Why don't electrons from the surroundings jump onto the positive dome?* Move a finger closer and closer to the charged dome, until a spark jumps between it and the dome. Demonstrate the plasma ball, if you have one.

Thunderclouds Ask: *What is lightning?* Explain how thunderclouds build up much larger charges and p.d.s between the clouds and the Earth, and so can pull electrons across large distances. We call these sparks lightning; the noise they make is thunder.

Revisit the questions from Trigger interest. Describe and explain examples of how sparks can be dangerous (refuelling planes, working with small electrical components) and what earthing is. Then describe how static electricity can be useful (such as spray-painting cars or photocopiers).

Main activity

Activity: Uses of static electricity (35 mins) Students draw a cartoon strip to explain how lightning happens, then answer questions about uses of static electricity.

Review and reflect

Ask students to discuss the questions on the Lesson presentation with their partner, then feed back to the class.

Homework

Ask students to find out what arc welding is.
(Two pieces of metal to be joined are melted by a large current, caused by holding a high-voltage electrode near the join. The arc is the spark that jumps from the electrode to the metal.)

Theme: Electricity and magnetism

Sub-themes: Circuits, Currents, resistance, and potential difference

Physics NC links

- electric current, measured in amperes, in circuits, series and parallel circuits, currents add where branches meet and current as flow of charge
- potential difference, measured in volts, battery and bulb ratings; resistance, measured in ohms, as the ratio of potential difference (p.d.) to current

Working scientifically NC link

- using a variety of concepts and models to develop scientific explanations and understanding (KS4)

Learning objective

- Explain the difference between current, potential difference, and resistance using models and ideas about fields.

Year 7 and 8 learning objectives to be applied

- Give the definition of an electric field
- Give the definition of current
- Give the definition of potential difference in terms of energy and force
- Give the definition of resistance

Application

- Conventional current.

Tier 2 vocabulary	Tier 3 vocabulary
	conventional current, electric field

SB in-text question answers

A If the electrons are moving faster, the current is bigger **B** Electrons are negatively charged **C** The mass would be higher up

Lesson plan

Reactivate knowledge

1 What is the definition of current?
2 What is the definition of potential difference?
3 How do you calculate resistance?

Lesson resources

Activity: *Circuit models and making a cell* (student sheet, support sheet, teacher and technician sheet)

Answers: 1 Current is the charge flowing per second **2** P.d. is the energy transferred per charge **3** Resistance = $\frac{\text{p.d.}}{\text{current}}$

Trigger interest

Ask: *Is there electricity 'in' the wires of a circuit? Can you 'count' electricity?* Discuss and explain that electricity is not a quantity, like current and charge. Discuss the importance of being precise with the words we use when talking about circuits.

Exposition of main content

Ask: *Do you know how a cell works?* Discuss, and explain that inside a cell there is a chemical reaction happening that separates molecules into positive and negative ions, which build up at opposite ends of the cell.

Explain that models help us understand things we can't see. Introduce the models for electric circuits on the Lesson presentation, and discuss the questions. (*What represents the cell? What represents the current? What represents resistance? What does the meter represent? What does the tap represent?*)

Main activity

Activity: Circuit models and making a cell (40 mins)
Working in groups, students study three models to describe electric circuits; for each, they decide what each part represents and describe how the model helps us understand these quantities.

They then make a voltaic pile and use it to light an LED. A voltaic pile is the very first type of cell ever invented.

As an extension, the rope model (described online by the Institute of Physics) can be considered as an alternative circuit model.

Review and reflect

Ask students to design a model to help younger children understand how the voltaic pile they made works.

Homework

Students research how a lemon cell works.

Theme: Electricity and magnetism

Sub-themes: Circuits, Current, resistance, and potential difference

Physics NC link

- electric current, measured in amperes, in circuits, series and parallel circuits, currents add where branches meet and current as flow of charge

Working scientifically NC link

- interpret observations and data, including identifying patterns and using observations, measurements and data to draw conclusions

Learning objective

- Describe some uses of series and parallel circuits, and how you can use them to sense the environment.

Year 7 and 8 learning objectives to be applied

- Describe the difference between series and parallel circuits
- Describe how current varies in series and parallel circuits
- Describe how potential difference varies in series and parallel circuits

Application

- Variable resistors and LDRs.

Tier 2 vocabulary	Tier 3 vocabulary
	light- dependent resistor (LDR), variable resistor

SB in-text question answers

A It decreases **B** 1.5 V/the p.d. across the battery **C** So that if one headlamp breaks while you are driving, you can still use the other one to see

Lesson plan

Reactivate knowledge

1 What is the difference between series and parallel circuits?
2 What can you say about the current everywhere in a series circuit?
3 What can you say about p.d. across single components in parallel?

Lesson resources

Practical: *Investigating a light dependent resistor (LDR)* (student sheet, support sheet, teacher and technician sheet)

Answers: 1 A series circuit contains one loop and a parallel circuit contains more than one loop **2** It is the same **3** It is the same

Trigger interest

Show students the images on the Lesson presentation. Ask: *How could you change the circuit to make the bulb dimmer?* Discuss ideas, such as using a lower-voltage cell or adding more components (which increases the resistance and so lowers the current). Point out the variable resistor and explain its function, and how it is the circuit's 'dimmer switch'. Introduce LDRs as another component whose resistance can change.

Exposition of main content

Show the image of an LDR on the Lesson presentation and ask: *How can we change the resistance of this component?* Describe how they work.

Remind students how a variable resistor is used in a series circuit to dim a bulb. Describe how it can also be used to control volume and on a mixing desk.

Show students the hairdryer circuit. Ask: *Why is a parallel circuit better than a series circuit in this example?* Guide discussion towards the fact that different parts of the circuit can be switched on and off separately.

Main activity

Practical: Investigating a light-dependent resistor (LDR) (35 mins) Students carry out a practical to determine whether the resistance of an LDR increases or decreases, when the brightness of the light falling on it increases. Then they plot a graph of LDR current against bulb p.d.

Review and reflect

Model, for students, how to describe and explain a pattern in as much detail as possible. Ask students to write a conclusion for their practical, relating the voltage across the bulb to the bulb brightness, the brightness of light hitting the LDR, the resistance of the LDR, and the current in the LDR circuit.

Homework

Tell students that LDRs are used in appliances where you want something to switch on or off automatically when the light levels rise or fall to a particular brightness. Ask students to think of appliances where something happens automatically because the light levels get high/low enough, for example, an alarm clock that goes off whenever the natural light in the room rises above a given level.

Theme: Electricity and magnetism

Sub-theme: Magnetism

Physics NC link

- exploring the magnetic fields of permanent and induced magnets, and the Earth's magnetic field, using a compass (KS4)

Working scientifically NC link

- using a variety of concepts and models to develop scientific explanations and understanding (KS4)

Learning objective

- Use the domain theory of magnetism to explain permanent and induced magnetism.

Year 7 and 8 learning objectives to be applied

- Give the definition of a magnetic field
- Describe how to investigate and represent the shape of a magnetic field

Application

- Domain theory.

Tier 2 vocabulary	Tier 3 vocabulary
demagnetise, domain, magnetise	*induce* induced magnet, permanent magnet

SB in-text question answers

A Aluminium **B** Yes, the rocks are behaving like magnets **C** You need to remove the magnet to see if the material is still magnetic

Lesson plan

Reactivate knowledge

1 Define 'magnetic field'.

2 Name 3 magnetic materials.

3 Describe how a compass works.

Lesson resources

Practical: *Investigating magnetism* (student sheet, support sheet, teacher and technician sheet)

Answers: 1 A region where a magnetic material or magnet experiences a force **2** Any three from iron, steel, cobalt, nickel **3** It contains a small magnet that lines up with the magnetic field of the Earth

Trigger interest

Ask: *Why can you magnetise a piece of iron but not a piece of glass? What do you know about the structure of metals?* Guide discussion to the fact that not all materials – including many metals – are magnetic, and that atoms in metals form a lattice structure.

Exposition of main content

Ask: *What is the domain theory of magnetism?* Introduce the idea that magnetic materials are magnetic because they are made up of small regions called domains. Domains behave like magnets.

Ask: *How do you magnetise or demagnetise materials?* Introduce the two ways a steel needle can be magnetised (stroking with a permanent magnet, or placing in a strong magnetic field). Explain that a material can be demagnetised by heating or hitting, as these both cause the domains to fall out of alignment again.

Ask: *Do you know what hard and soft magnetic materials are? What is the difference between permanent and induced magnets?* Guide discussion towards the correct answers. Take a moment to teach the meaning of 'induce'. Note that both types of magnet are useful. Students will see why soft magnetic materials are useful in the next lesson.

Main activity

Practical: Investigating magnetism (20 mins)
Students make a floating compass by stroking a needle with a permanent magnet and then investigate hard and soft magnetic materials.

Review and reflect

Ask students to write quiz questions about magnetism, then test a partner – they should decide if answers are worth one, two, or three marks. Ask what students decided were the 'three-point answers' and why, then discuss what strategies they could use to remember them.

Homework

Ask students to find out what lodestones are, and what they were used for.

Theme: Electricity and magnetism

Sub-theme: Electromagnetism

Physics NC links

- the magnetic effect of a current, electromagnets, D.C. motors (principles only) (KS3)
- magnetic effects of currents, how solenoids enhance the effect (KS4)

Working scientifically NC link

- ask questions and develop a line of enquiry based on observations of the real world, alongside prior knowledge and experience

Learning objectives

- Describe how to induce a p.d.
- Describe what affects the sign and magnitude of the p.d.
- Describe how a simple generator works.

Year 7 and 8 learning objectives to be built on

- Describe how to make an electromagnet
- Describe how a simple motor works

Key Stage 4 link

- Electromagnetic induction.

Tier 2 vocabulary	Tier 3 vocabulary
	electromagnetic induction, electromagnetism, *induce*

SB in-text question answers

A Make the current bigger **B** Voltmeter **C** The p.d. would be negative/opposite sign

Lesson plan

Reactivate knowledge

1 Describe the shape of the magnetic field around a current-carrying wire.

2 Describe the three main parts of a thermal power station.

3 Describe how to make a simple motor.

Answers: 1 Circular/concentric circles **2** For example, boiler, turbine, generator, cooling towers **3** Put a current-carrying coil of wire in a magnetic field

Lesson resources

Practical: *Investigating inducing a p.d. and investigating generators* (student sheet, support sheet, teacher and technician sheet)

Trigger interest

Ask students: *What connects electricity and magnetism?* Discuss similarities between electric forces and magnetic forces, such as two signs/poles, opposites attract and like repels, non-contact forces. Note that this is no coincidence; the connection between electricity and magnetism is fundamental. In fact, we talk about electromagnetism, which is the interaction between electric charges and electric and magnetic fields.

Exposition of main content

Magnetic field around a current-carrying wire Use a compass to demonstrate the magnetic field around a current-carrying wire. Say that you have induced a magnetic field. Check understanding of the word 'induce' again.

Magnetic field around an electromagnet Demonstrate an electromagnet and describe the shape of the field around it. Link the role of the iron core (soft magnetic material) to the previous lesson.

Inducing a potential difference Demonstrate inducing a p.d. using the measuring cylinder apparatus. Note that it doesn't matter if the magnet moves through the coil or if the coil moves past the magnet; what matters is that they are moving past each other. Discuss students' observations about how the induced p.d. is related to the motion of the magnet. Tell students they will investigate this further. Discuss what 'detailed observations' are.

Main activity

Activity: Investigating inducing a p.d. and investigating generators (25 mins) In Task 1, students experiment with the measuring cylinder apparatus and write down as many and as detailed observations as they can on how the induced p.d. is related to the position and movement of the magnet.

In Task 2, if the equipment is available, students investigate simple generators to test factors that increase the size of the induced current. Alternatively, students answer the questions on the worksheet.

Review and reflect

Ask students to write their observations on post-it notes and stick them on the board, grouping similar observations together. Select observations to read out; other students can suggest what is good about the observation and how they made them.

Homework

Ask students to find three products that use portable generators.

Theme: Electricity and magnetism

Sub-themes: Curren, resistance, and potential difference, Domestic electricity

Physics NC link

- how transformers are used in the national grid and the reasons for their use (KS4)

Working scientifically NC links

- explaining everyday and technological applications of science; evaluating associated personal, social, economic, and environmental implications; and making decisions based on the evaluation of evidence and arguments (KS4)
- interpret observations and data, including identifying patterns and using observations, measurements and data to draw conclusions (KS3)

Learning objectives

- Describe the difference between alternating and direct current
- Describe how a.c. is used to induce a p.d.
- Describe what the National Grid is.

Year 7 and 8 learning objectives to be built on

- Describe some uses of electromagnets
- Describe how electricity is generated with renewable and non-renewable resources

Key Stage 4 link

- AC/DC, transformers, and the National Grid.

Tier 2 vocabulary	Tier 3 vocabulary
AC, DC, National Grid	alternating current, direct current, *induction*, transformer

SB in-text question answers

A The peaks would be closer together/the wavelength would be shorter **B** A p.d. is induced in the metal pan, inducing a current
C 50 Hz

Lesson plan

Reactivate knowledge

1 What is the difference between current and p.d.?
2 Define frequency.
3 Describe the field around an electromagnet.

Lesson resources

Practical: *Transformers* (student sheet, support sheet, teacher and technician sheet)

Answers: 1 Current is the charge flowing per second and p.d. is the energy per charge **2** Number of waves/cycles per second **3** It is the same shape as that around a bar magnet

Trigger interest

Ask: *Have you ever wondered where the electrical sockets in your house lead to? How does the p.d. produced in a power station get to your house? Do you know how many volts of p.d. are supplied to British houses?*

During discussion, write students' ideas on the board, until you reach an overview of the National Grid, as shown on the Lesson presentation.

Exposition of main content

AC/DC Ask students: *Do you know what DC and AC are?* Then ask: *Does the National Grid supply AC or DC?* Introduce what AC and DC are; show the two graphs on the Lesson presentation. Draw a few irregular AC and DC graphs on the board to test students' understanding (e.g., a sinusoidal shape that lies entirely above the time axis (still DC) or a single, straight diagonal line crossing the time axis (AC)).

Transformers Ask: *Do you know what a transformer is?* Introduce the parts of a transformer, using the diagram on the Lesson presentation. Point out that the primary coil is an electromagnet. Explain how the AC through the primary coil leads to a continually changing magnetic field that extends through the secondary coil, which induces another AC in the secondary coil. Note that having different numbers of turns on the primary and secondary coils changes the p.d. across and current through the secondary coil – students will investigate this later.

The National Grid Explain what is meant by 'stepping up' and 'stepping down' current and p.d., and why stepping down current immediately after it leaves the power station is an advantage for the National Grid. Note why this means that power lines need to be very high up.

Main activity

Practical: Transformers (25 mins) Keeping the primary AC and p.d. the same, students vary the number of turns on the secondary coil and record the resulting secondary current and p.d. .

Review and reflect

Ask students to work with a partner to plan an answer to the exam-style question shown on the Lesson presentation. As a class, discuss content and structure.

Homework

Students write their answer to the exam-style question.

Introduction to checkpoint intervention

This checkpoint intervention provides suggestions for a lesson to follow up the P3 4 *Electricity and magnetism* checkpoint assessment. Depending on the outcomes of the assessment, these suggestions could move a class towards secure, to target specific learning outcomes, or to consolidate knowledge for students achieving secure outcomes. Use students' outcomes from the checkpoint assessment to address general misconceptions from the content covered. Be prepared to re-cover content as required; three key concepts from the chapter are covered in more detail below.

Checkpoint secure learning outcomes

- Explain why sparks happen.
- Describe how static electricity can cause problems and how it can be useful.
- Explain the difference between current, potential difference, and resistance using models and ideas about fields.
- Describe some uses of series and parallel circuits, and how you can use them to sense the environment.
- Use the domain theory of magnetism to explain permanent and induced magnetism.

Student reflection

Encourage students to reflect on whether there were any checkpoint questions they found difficult or straightforward, as well as their preparation for the checkpoint assessment, including how they revised, the time spent on revision, and what areas they could improve on. For more strategies, see the Metacognition chapter or the Metacognition in Key Stage 3 Science guide on Kerboodle.

Intervention activity: Explain why sparks happen

The idea of an electric field between two points that have different potentials is challenging, as students cannot *see* it. However, it is possible to see the spark between objects, so the idea of charging something up to make it more likely for a spark to jump from it to something else is accessible.

Van der Graaf

Use a Van der Graaf generator to demonstrate that charging the dome for longer allows charged particles to jump a larger gap, leading to a bigger spark. Stress that this is because a *stronger electric field exists between the dome and the discharge ball.*

Intervention activity: Explain the difference between current, potential difference, and resistance using models and ideas about fields

Potential difference and 'per unit charge' are difficult concepts. Students are used to the term 'rate' involving time and/or money, but not other quantities. Students also have difficulty understanding why more energy per unit charge results in the charged particles being pushed faster round the circuit.

Cross-country circuit model

Present another model for a circuit: students running a cross-country race. Energy is represented by chocolate. The idea is that the more energy the students are given, the faster they will run. Resistance is represented by how steep different hills are. You may choose to represent charge by the students' ages, as follows. All students are 13 years old. This represents them all having the same charge. They are given 13 pieces of chocolate, when they go through the 'cell'. Replacing the cell with one that has twice as much voltage means that they all get given 26 pieces of chocolate as they pass through it instead, and so they run faster.

Intervention activity: Describe some uses of series and parallel circuits, and how you can use them to sense the environment

Many students find the difference between series and parallel circuits difficult to grasp, perhaps because they do not distinguish between the section of the circuit with the cell in it and the other sections that only have components in them.

Walk-around circuits

Draw some series and some parallel circuits outside. Ask students to walk around them. Instruct them to go through the cell, but assign which branch of each parallel circuit each student should go through.

Biology Chapter 1: Cells

Biology Chapter 1 Opener

Reactivate your knowledge questions

1. Red blood cell
2. Controls what moves in and out of the cell
3. glucose + oxygen → carbon dioxide + water (+ energy)

Biology Chapter 1.2: Apply: Plant and animal cells

Summary questions

1. Nucleus, cell, mitochondria, chloroplasts, wall, vacuole, microscope [7 marks]
2. **a)** Control what substances come in and out of the cell [1 mark]
   b) Nucleus/cytoplasm/mitochondria [1 mark]
   c) Any one from: [1 mark for component, 1 mark for description]
   vacuole – contains a watery fluid/keeps the cell firm
   cell wall – strengthens the cell/provides support
   chloroplasts – contain chlorophyll/where photosynthesis occurs.
3. Any three from:
   As microscopes develop, they become more powerful
   This allows scientists to observe objects under a higher magnification/better resolution
   This allows more details to be seen within cellular components
   This allows structures to be seen in 3D
   This may allow scientists to see the movement of substances within/into/out of a cell.
4. Any four from:
   Cell wall contains cellulose
   This strengthens the cell/provides support
   Plant cells are arranged in a regular structure/fit together like bricks.
   Cell wall contains cell sap which keeps the cell firm
   Cell sap/liquid presses against the cell wall to provide rigidity.
5. Any six from:
   Only plants contain a cell wall, vacuole, and chloroplasts
   They need a cell wall and a vacuole for structure as they do not have a skeleton
   cells are also arranged in a more regular structure than animal cells
   allowing the cells to fit together like bricks
   a large permanent vacuole filled with cell sap keeps the cell firm
   cell wall made of tough cellulose, this strengthens the cell/provides support.
   chloroplasts, packed with chlorophyll trap light for photosynthesis
   animals do not need chloroplasts as they take in glucose when they eat.
6. Any six from: [To award six marks, at least two relevant examples must be given]
   Before the invention of the microscope cells had not been discovered/cells are too small to be seen with a naked eye.
   Light microscopes allowed Hooke to see the first cells to be seen.
   At this time no internal components of a cell were known.
   Van Leeuwenhoek observed the first living cells/saw structures inside cell.
   Structures like chloroplasts and nuclei can be seen using a modern light microscope.
   Stains/named stain can be used to make cell components easier to see.
   By studying lots of samples scientists (Schleiden and Schwann) proposed cell theory/that all living organisms are made of cells.
   Electron microscopes can magnify images up to several million times
   allowing the structures inside cell components to be seen/such as the internal structure of a chloroplast/other suitable example.
   3D images of cell component structures due to further developments in electron microscopy.

Biology Chapter 1.3: Apply: Cell specialisation

Summary questions

1. Specialised, xylem, phloem, sieve, food/sugars [5 marks]
2. **a)** Phloem [1 mark]
   b) Transport water [1 mark] and minerals. [1 mark]
   c) Cells die to form hollow tubes [1 mark]. Lignin is very strong/supports the plant stem. [1 mark]
3. **a)** Both form tubes [1 mark] both transport materials around the plant. [1 mark]
   b) Any three from:
   Xylem cells are dead, phloem cells are alive
   Phloem cells have companion cells, xylem cells do not
   Xylem cells have lignin, phloem do not
   Phloem cells have sieve plates between the cells, xylem cell walls disappear completely.
4. Unicellular organisms consist of only one cell, [1 mark] therefore this cell must perform all the processes needed to sustain life, [1 mark] specialised cells are adapted to form a particular function. [1 mark]
5. Any four from:
   Contain proteins that slide over each other
   to make the fibres contract/shorten
   Have many mitochondria
   to transfer the energy needed for the proteins to move
   Store glycogen
   which can be broken down into glucose for respiration

6. Any six from:
Cell walls between the phloem cells break down to form sieve plates/tubes,
allowing water containing dissolved food to move freely along the tube.
They are supported by companion cells
which help keep the phloem cells alive
as much of the phloem cells' internal structure has been lost (to maximise space).
Companion cells contain mitochondria to transfer the energy needed to move the dissolved food.

Biology Chapter 1.4: Apply: Respiration

Summary questions

1. Rate, breathing, muscle, blood, respiration [5 marks]
2. Glucose and oxygen react [1 mark] to form carbon dioxide and water; energy is transferred to the cells. [1 mark]
3. To transfer more energy [1 mark] so the cell can contract faster/harder/do more work. [1 mark]
4. **a)** It increases. [1 mark]
   b) To pump blood to the muscle cells faster [1 mark] so the cells can respire more. [1 mark]
5. **a)** 65 (±2) breaths per minute [1 mark]
   b) 65 (±2) − 15 = 50 (±2) breaths per minute [2 marks, allow identification of 15 for 1 mark]
   c) So more oxygen is taken into the lungs [1 mark] and transported to the muscle cells for respiration[1 mark] and more carbon dioxide can be removed. 1 mark]
6. Any six from:
Breathing rate increases
so more oxygen is taken into the lungs
(and) so more carbon dioxide is removed.
Heart rate increases
so blood containing oxygen and glucose is transported to the muscles faster
(and) to increase the rate at which carbon dioxide is removed from the muscles.
Glycogen stored in the muscle is broken down/releases glucose.
More oxygen *and* more glucose allow cells to respire *more*/at a greater rate
transferring *more* energy to the cells.

Biology Chapter 1.5: Apply: Diffusion

Summary questions

1. Rate, temperature, area, gradient [4 marks]
2. **a)** Movement of water particles from an area of high (water) concentration to a low (water) concentration [1 mark] through a partially permeable membrane. [1 mark]
   b) Movement of water into root hair cells from the soil/ movement of water between root hair cells and the xylem vessels (accept other suitable examples). [1 mark]
3. The higher the temperature, the greater the rate of diffusion [1 mark] as the particles are moving faster. [1 mark]
4. Dye droplet has a high concentration of coloured particles, [1 mark] water has a low concentration of coloured particles. [1 mark] Coloured particles move from an area of high concentration to low concentration [1 mark] until they are evenly spaced throughout the water/they are the same concentration throughout the liquid. [1 mark]
5. **a)** Villi increase the surface area/provide a larger surface area) [1 mark] This increases the rate of diffusion as more particles can move across the membrane at the same time. [1 mark]
   b) Any three from:
   This lowers the concentration of glucose in the blood at the villi
   so the concentration of glucose is lower in the blood than in the small intestine.
   This maintains a steep/high/large concentration gradient.
   The larger the concentration gradient, the higher the rate of diffusion.
6. Any six from:
There are millions of alveoli in the lungs
with curved surfaces in contact with the bloodstream/ capillaries.
This creates a large surface area.
The larger the surface area, the greater the rate of diffusion (of oxygen into the blood)
as there is more space so more particles can move at the same time.
The alveoli have very thin walls/one cell thick.
This creates a short diffusion distance.
As the blood transports oxygen away to the rest of the body
it maintains a steep/large concentration gradient (of oxygen in the alveoli compared to the blood).
The steeper the concentration gradient, the faster the rate of diffusion.

Biology Chapter 1.6: Reach: Prokaryotic cells

Summary questions

1. [Allow 1 mark for each correct row, max 7]

Component	Pl	An	Pr
Cell membrane	✓	✓	✓
Cell wall	✓		✓
Chloroplast	✓		
Cytoplasm	✓	✓	✓
Mitochondria	✓	✓	
Nucleus	✓	✓	
Permanent Vacuole	✓		

2. In a eukaryotic cell genetic material is found in the nucleus whereas in a prokarytotic cell genetic material is found free floating in the cytoplasm. [1 mark] DNA is arranged into many chromosomes/strands in a eukaryotic cell whereas in a prokaryotic it is one long (circular) strand/ chromosome. [1 mark] Prokaryotic cells (can) contain extra rings of DNA/plasmids. [1 mark]
3. Any six from:
Both have cytoplasm *and* a cell membrane.
Both have a cell wall,
in plants it is made of cellulose but in a prokaryotic cell it is made of peptidoglycan.
Both have genetic material/DNA,
in plants the genetic material is in the nucleus but in prokaryotes it is in the cytoplasm.
Only plants contain mitochondria/chloroplasts.
Only prokaryotic cells have a slime capsule/pili/flagella.

Biology Chapter 1.7: Reach: Active transport

Summary questions

1. Low, high, gradient, energy [4 marks]
2. Active transport requires energy, diffusion doesn't. [1 mark] In diffusion, particles move from a high to a low concentration, while in active transport particles move from a low to a high concentration. [1 mark]
3. Any six from:
When the concentration of glucose is lower in the blood than the intestine, glucose moves by diffusion.
The glucose molecules move from a high concentration to a low concentration with/down the concentration gradient.
This is a passive process/does not require energy.
When the concentration of glucose is higher in the blood than the intestine, glucose moves by active transport.
The glucose molecules move from a low concentration to a high concentration against/up the concentration gradient.
This process requires energy.

Biology Chapter 1 Summary Answers

1. a) [Award 3 marks if all 4 correct, 2 marks if 2 are correct, 1 mark if 1 is correct]
Cytoplasm – where chemical reactions take place
Chloroplast – where photosynthesis occurs
Cell membrane – controls what comes in and out of the cell
Nucleus – controls the cell and contains DNA
b) (Permanent) vacuole/cell wall [1 mark]
c) Vacuole – filled with cell sap, keeping the cell firm
or
cell wall – made of cellulose, strengthens the cell/provides support. [2 marks]
2. Xylem cell [1 mark]
b) Cell C [1 mark]
c) Transfer male genetic material to the egg. [1 mark]
d) Any two from: no nucleus/contains haemoglobin/ biconcave disc [2 marks]
3. a) 15 [2 marks; allow 1 mark for (13+ 17 + 15)/3]
b) Exercise causes an increase in the breathing rate [1 mark] and heart rate [1 mark].
c) Any four from:
Exercise requires energy [1 mark] to be transferred more rapidly to (muscle) cells [1 mark] so blood containing oxygen [1 mark] and glucose [1 mark] is transported to the muscles more quickly [1 mark] to allow an increase in the rate of respiration. [1 mark]
4. a) [1 mark for each correctly completed row]

Feature	Xylem vessels	Phloem vessels
substances carried	Water (and minerals)	Dissolved sugar / food
direction of transport	From roots to leaves / up the plant	From leaves all over the plant / up and down
contains living cells	No	Yes
description of cell walls between cells	Missing	Sieve plates
lignin present?	Yes	No

b) Companion cells contain mitochondria [1 mark] to transfer the energy needed to move the dissolved food. [1 mark].
5. Any six from:
Light microscopes allowed the first cells to be seen.
When cells were first discovered no internal components of a cell were known
Structures like nuclei/cell membrane/cytoplasm can be seen using a modern light microscope.
Stains/named stain can be used to make cell components easier to see.
Stain, e.g., methylene blue, is used to colour the nuclei.
Electron microscopes can magnify images up to several million times.
This allows the structures inside cell components to be seen
such as the internal structure of a mitochondrion.
3D images of cell component structures have been produced due to further developments in electron microscopy
6. a) Bacteria [1 mark]
b) Any two from: contain a cell membrane/cytoplasm/ genetic material. [2 marks; accept cell wall if reference is to a eukaryotic plant cell]

c) [Award 1 mark for a correct difference and 1 mark for a linked description]
In eukaryotic cells, the genetic material is in the nucleus, but in prokaryotes it is in the cytoplasm/ prokaryotic cells walls are made from peptidoglycan, but eukaryotic plant cell walls are made from cellulose.

7. a) B Particles move against a concentration gradient [1 mark], C The process requires energy. [1 mark]
   b) Suitable example, e.g., uptake of mineral (ions) into a root/root hair cell, uptake of glucose into the small intestine. [1 mark]

Biology Chapter 2: Cell Systems

Biology Chapter 2 Opener

Reactivate your knowledge questions

1. Movement, nutrition, sensitivity, reproduction, growth, respiration, excretion
2. Transfer of energy from organic molecules to your cells
3. Allow gases to diffuse in and out of the leaf

Biology Chapter 2.2: Apply: Cells to organ systems

Summary questions

1. Cell, tissue, organ, organ system, organism [5 marks]
2. Sperm, muscle tissue, lung, digestive system [3 marks if all correct, award 1 mark for identifying sperm as smallest structure or digestive system as the largest, award 2 marks for two correctly placed structures]
3. Compared to tissues, organ systems are larger structures [1 mark] and more complex. [1 mark]
4. a) Any two from:
   Observations of external features
   Experiments
   Dissections.
   Accept other suitable suggestions, e.g., chemical tests
   b) For example, technology/microscopes/named equipment/improve over time [1 mark] which allow internal structures to be seen in more detail. [1 mark]
   Accept other reasonable arguments/suggestions with reasoning.
5. Any six from:
   Cells are the first level of organisation/building blocks of life.
   Examples include muscle cells/red blood cells/white blood cells.
   Tissues are groups of cells working together to perform a function.
   Examples include blood/nervous tissue/cardiac muscle tissue.
   Organs are groups of tissues working together to perform a function.
   Examples include heart/blood vessels.
   Organ systems are groups of organs working together to perform a function.
   Circulatory system is an example of an organ system which transports substances around the body.
6. Any six from:
   Cells are the first level of organisation/building blocks of life.
   Examples include muscles/epithelial cells.
   Tissues are groups of cells working together to perform a function.
   Examples include (smooth) muscle tissue/glandular tissue.
   Organs are groups of tissues working together to perform a function.
   Examples include stomach/liver/pancreas/intestine.
   Organ systems are groups of organs working together to perform a function.
   Digestive system is an example of an organ system which breaks down food and absorbs nutrients.

Biology Chapter 2.3: Apply: Digestive system and enzymes

Summary questions

1. Catalysts, speed, used, digestive, baby, washing [6 marks]
2. a) Carbohydrates [1 mark] to sugars. [1 mark]
   b) Lipids [1 mark] to fatty acids and glycerol. [1 mark]
3. Proteases [1 mark] break down proteins into amino acids [1 mark] which are easier for babies to digest/babies cannot digest proteins. [1 mark]
4. Grease stain is caused by fat. [1 mark] Lipases [1 mark] break down lipids into fatty acids and glycerol. [1 mark] These dissolve in the water/are water soluble/can be washed away. [1 mark]
5. a) Biological – contains enzymes [1 mark]
   b) Any four from:
   Enzymes digest insoluble stain molecules into (smaller) soluble molecules which will dissolve easily/at low temperatures.
   Protease molecules in the detergent may damage protein fibres in silk/wool.
   Some people are allergic to enzymes which may remain on clothing after washing, causing skin irritation.
6. Any six from: [Answer must contain at least one advantage and one disadvantage to gain full marks]
   Digestive enzymes break down stains [Credit example of named enzyme and its use]
   Advantages:
   Enzymes work at relatively low temperatures
   Energy /money are saved by allowing clothes to be washed at a lower temperature
   Clean delicate fabrics which would otherwise be damaged by a hot wash.
   Disadvantages:
   Can cause allergic or other reactions
   Protease enzymes may damage protein fibres in fabrics such as silk/wool.

Biology Chapter 2.4: Apply: Respiratory system and gas exchange

Summary questions

1. Exchange, environment, villi, alveoli [4 marks]
2. a) Provide a large surface area. [1 mark]
   b) Rich blood supply/thin/one cell thick. [1 mark]
3. Any four from.
   Multicellular organisms have a small surface area to volume ratio.
   Diffusion does not occur fast enough (over a large distance) to sustain life/meet cells' demands (so they rely on exchange surfaces).
   Unicellular organisms have a large surface area to volume ratio.
   Everything they need can diffuse through their surface quickly enough to sustain life as diffusion distances are small.
4. Any six from:
   Thin alveoli wall
   and short diffusion distance
   increases rate of diffusion of oxygen into the blood/carbon dioxide out of the blood.
   Lots of them/spherical shape
   increases surface area
   maximising number of particles which can move at the same time.
   Good blood supply/ventilation
   removes carbon dioxide/brings in oxygen (or vice versa)
   maintaining steep concentration gradient between gas in the lungs and the blood
   maximising rate of diffusion.
5. Any six from: [as it is a comparison question award a maximum of 5 if the difference is not mentioned]
   Both have a good blood supply
   maintaining concentration gradient between blood and lungs/blood and intestine,
   maximising rate of diffusion.
   Both have thin walls/are one cell thick
   and the short diffusion distance
   increases rate of diffusion of oxygen into the blood/carbon dioxide out of the blood.
   Both increase the surface area
   by their shape/having lots of them present,
   maximising number of particles which can move at the same time.

Biology Chapter 2.5: Apply: Leaf structure and photosynthesis

Summary questions

1. Stomata – allow gases to diffuse in and out of the leaf, waxy layer – reduces amount of water evaporating from the leaf, guard cells – open and close stomata, veins – transport water to cells in leaf, palisade mesophyll – main site of photosynthesis [Allow 3 marks for all correct, 2 marks for 3 or 4 correct, 1 mark for 1 or 2 correct]
2. a) Carbon dioxide [1 mark], water. [1 mark]
   b) Chloroplasts [1 mark]
   c) Contains xylem and phloem vessels, [1 mark] xylem vessels transport water (to the leaf), [1 mark] phloem vessels transport glucose/sugars (all over the plant). [1 mark]
3. Flat – creates a large surface area [1 mark] to maximise light absorption. [1 mark] Thin - provides a short diffusion distance [1 mark] for carbon dioxide to diffuse into the leaves. [1 mark]
4. Palisade mesophyll is the main site of photosynthesis. [1 mark] Palisade cells are column-shaped/packed with many chloroplasts [1 mark] to maximise light absorption. [1 mark] Spongy mesophyll is where gas exchange occurs. [1 mark] Cells are loosely packed/air spaces between the cells, [1 mark] allowing carbon dioxide to diffuse throughout the leaf/oxygen to diffuse out of the leaf. [1 mark]
5. Any six from:
   They contain chlorophyll which traps light.
   They are thin, providing a short diffusion distance.
   They have large surface area to maximise light absorption.
   They have veins/xylem vessels to bring water to the cells.
   Palisade mesophyll has cells packed with chloroplasts/near the surface of the leaf to maximise light absorption.
   Spongy mesophyll has air spaces to allow gases to diffuse through the leaf.
   Stomata allow gases to diffuse into and out of the leaf.
   Waxy layer prevents water evaporating.
6. Any six from: [accept converse for marking points below]
   Leaves growing in the shade are likely to be larger.
   The increased surface area
   increases amount of light which can be absorbed.
   Leaves growing in full sunlight are likely to have a thicker waxy layer
   to prevent water loss by evaporation.
   Leaves growing in full sunlight are likely to close their stomata more frequently
   to prevent water loss by evaporation.
   Accept any other plausible suggestions, e.g., darker green in the shade as it may suggest a higher chlorophyll content to maximise the rate of photosynthesis.

Biology Chapter 2.6: Reach: Circulatory system

Summary questions

1. Capillaries, arteries, veins, valves [4 marks]
2. Blood passes through the heart twice on each circuit of the body. [1 mark] One loop travels from the heart to the lungs and back, [1 mark] one loop travels from the heart to the rest of the body and back. [1 mark]
3. Any six from:
   Arteries carry blood away from the heart.
   They have thick walls to withstand high pressures.

These walls are elastic to stretch as blood passes.
Veins carry blood back to the heart.
They have valves to prevent backflow.
The valves have thinner walls as blood is at a lower pressure.
Capillaries form a network joining arteries to veins.
One cell thick so substances can diffuse easily between them.

Biology Chapter 2.7: Reach: Transpiration

Summary questions

1. Transpiration, roots, xylem, stomata [4 marks]
2. Any four from:
   In good photosynthesis conditions/high light levels, stomata open
   and guard cells takes up water (by osmosis), making them swell up (opening the stomata).
   When photosynthesis conditions are not good, stomata close
   and guard cell lose water,
   making them floppy (and the stomata close).
3. Any six from:
   (When stomata are open,) water evaporates from the leaf's cells to the leaf's air spaces.
   Higher concentration of water vapour in the leaf than in the surrounding air/concentration gradient between leaf and air,
   so water vapour diffuses out of the leaf/through stomata.
   Loss of water from the leaves reduces the pressure in the top of the xylem vessels.
   Water flows through the tube from high pressure (at the bottom) to lower pressure (at the top).
   Water diffuses into root hair cells
   as lower concentration of water in root cells (as water drawn up the transpiration stream).

Biology Chapter 2 Summary Answers

1. **a)** Tissue [1 mark]
   b) Group of organs working together to perform a function. [1 mark]
   c) Stomach/small intestine/large intestine/pancreas/mouth/gullet. [1 mark]
2. **a)**

Enzyme	Nutrient broken down	Product
Carbohydrases	Carbohydrate	**Sugars**
Proteases	Protein	Amino acids
Lipases	**Lipids/fats**	Fatty acids and glycerol

   b) Washing detergent/baby food/accept any other appropriate examples. [1 mark]
3. **a)** Spongy layer [1 mark]
   b) Main site of photosynthesis. [1 mark]
   c) Diffuses [1 mark] through the stomata [1 mark] into the air spaces/spongy mesophyll [1 mark] and then into the leaf cell/palisade cell. [1 mark]
4. **a)** Protease/lipase [1 mark]
   b) Protease breaks down protein [1 mark] such as blood/egg/sweat stains [1 mark] *or* lipase breaks down lipids [1 mark] such as oil/butter/fat/cheese. [1 mark]
   c) Some people experience allergic reactions/protease enzymes can damage protein fibres in natural fabrics/silk/wool. [1 mark]
   d) Enzymes work at relatively low temperatures/can remove stains at low temperatures, [1 mark] saves energy/money. [1 mark]
5. **a)** Volume = $1 \times 5 \times 20 = 100\ mm^3$ [1 mark]
   Surface Area: Volume = 250 : 100 [1 mark] = 2.5 : 1 [1 mark]
   b) It has a large surface area to volume ratio [1 mark] so that everything they need can diffuse through their surface quickly enough to sustain life [1 mark] as diffusion distances are small. [1 mark]
6. Any six from:
   Cells are the first level of organisation/building blocks of life.
   Examples include muscles/bone cells.
   Tissues are groups of cells working together to perform a function.
   Examples include muscle tissue/connective tissue.
   Organs are groups of tissues working together to perform a function.
   Examples include skull/ribs/biceps/other named bone/muscle group.
   Organ systems are groups of organs working together to perform a function.
   The muscular skeletal system is an example of an organ system which causes movement.
7. **a)** A It carries blood away from the heart. [1 mark]
   D It has a thick elastic and muscular wall. [1 mark]
   b) They have valves [1 mark] and thin walls/large lumen. [1 mark]
8. **a)** Xylem vessels [1 mark]
   b) Osmosis/diffusion [1 mark]
   c) Diffuses/moves out of the stomata) [1 mark]

Biology Chapter 3: Fertilisation and implantation

Biology Chapter 3 Opener

Reactivate your knowledge questions

1.
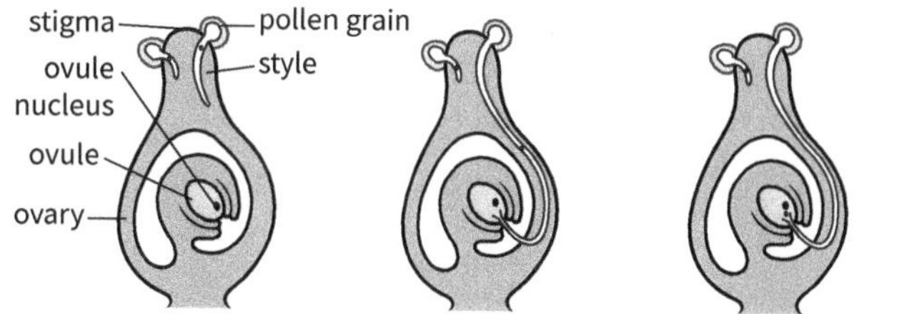

The tube grows out of the pollen grain and down through the style.

The pollen nucleus moves down the tube.

The pollen nucleus joins with the ovule nucleus. Fertilisation takes place and a seed will form.

2. Lining of the uterus
3. Condom might be pierced, inhibiting barrier method; spermicides might have separated from condom surface, inhibiting chemical method.

Biology Chapter 3.2: Apply: Sexual reproduction

Summary questions

1. Two, sexual, gametes/sex cells, fertilisation, variation [5 marks]
2. Joining together of the nuclei [1 mark] of a sperm and egg cell. [1 mark]
3. Two parents are needed, [1 mark] genetic material from the parents is mixed (during fertilisation), [1 mark] variation present in the offspring. [1 mark]
4. Any six from:
   both involve gametes
   both require two parents
   both involve the mixing of genetic material
   both produce variation in the offspring
   in both, the ovaries produce the female gametes
   in plants the male sex cell is pollen, in animals it is sperm
   in plants, the anther produces the male gamete, in animals it is the testes
   in plants, fertilisation occurs in the ovary, in animals it occurs in the fallopian tube/uterus.
5. They inherit half their genetic material from their mother, [1 mark] which will cause them to display some maternal characteristics, [1 mark] and half their genetic material from their father. [1 mark] This combines and mixes during fertilization. [1 mark]

Biology Chapter 3.3: Apply: Preventing pregnancy

Reading graphs

Condom – 3, contraceptive pill – 1, contraceptive implant – 1, spermicides – 28, no contraceptive – 85

Summary questions

1. Contraception, barrier, hormonal, kill [4 marks]
2. Act as a barrier [1 mark] which prevents sperm coming into contact with the egg. [1 mark]
3. Condoms are more effective than spermicides, [1 mark] they also protect against STIs. [1 mark]
4. **a)** Spermicides (no protection is not contraception). [1 mark]
   b) Any three from:
   Condoms have a very high effectiveness.
   Hormonal methods have a very high effectiveness.
   Spermicides are less effective.
   Both methods of contraception are many times more effective at preventing pregnancy than not using contraception.
5. Any two reasons and two linked explanations [4 marks]:
   prevent ovulation – stops the egg being released so it can't join with the sperm
   make the mucus in the cervix thick – stops sperm getting through to meet the egg
   prevent the uterus lining developing – so a fertilised egg cannot implant in the lining and develop into a baby
6. Any six from (accept the converse for each point):
   Both have a high degree of effectiveness
   Condom has no side effects
   Condom requires no medical approval
   Condom protects against STIs
   Pill may have side effects
   Pill has to be taken every day
   Condom may reduce feelings of pleasure.
   Accept any other relevant point.

Biology Chapter 3.4: Apply: The menstrual cycle

Interpreting graphs

The oestrogen level increases from day 0 to day 14, reaching a peak at day 14, and then decreases from day 15 to day 28.

Summary questions

1. Period, uterus, ovulation, fertilised, menstrual [5 marks]
2. **a)** 28 days [1 mark]
   b) Thick/at its thickest point. [1 mark]
   c) Day 5 [1 mark]
3. Oestrogen causes the uterus lining to build up, [1 mark] LH causes ovulation/egg to be released. [1 mark]
4. **a)** Uterus is thickest in range 22–24 days and thinnest in range 2–4 days, [1 mark] so roughly 20 days. [1 mark]
   b) The uterus lining increases [1 mark] to allow a fertilised egg to implant. [1 mark] If the egg does not implant, the old lining is lost. [1 mark]
5. Day 9 (12–3 days) [1 mark] to day 18 (16+2 days). [1 mark]
6. Any six from:
   From day 0 – day 14, oestrogen levels rise,
   causing the lining of the uterus to build up.
   Oestrogen levels then begin to fall/fall for the rest of the cycle,
   resulting in the lining of the uterus breaking down and being lost at day 28.
   LH levels rise to a peak in the middle of the cycles/day 14.
   At its peak, it stimulates the release of an egg/ovulation,
   the levels of LH then fall at the rest of the cycle
   (so no more eggs are released).

Biology Chapter 3.5: Apply: Investigating seed dispersal mechanisms

Summary questions

1. Dispersed, competition, valid, variables [4 marks]
2. Bright and tasty – so animals eat them, transporting them away from parent plant. [1 mark] indigestible seed coat – so the seed does not get damaged while inside the animal. [1 mark]

3. Release from same/named height each time, [1 mark] use seeds of same wing length / surface area, [1 mark] expose seeds to same/no wind speed each time. [1 mark]
4. Accept any two relevant suggestions, with explanations. Examples could include: [1 mark for suggestion, 1 mark for explanation]
   Would not be possible to blow bags up to same pressure/ equivalent each time, so they may have a different explosive force when popped.
   It would not be possible to use the same force each time to pop the bags, so some seeds may receive a larger/ smaller explosive force.
   Measuring the distance travelled by one seed may not reflect the mean distance travelled, so the data may not reflect the mean distance travelled/there may be significant outliers/random errors within the data.
5. **a)** 39mm wing length [1 mark]
   c) 2.5s (accept answer in range 2.4s to 2.6s) [1 mark for extrapolating backwards method, 1 mark for answer]
6. Any six from:
   No repeat results were collected,
   so there might be some (random) errors in the data.
   From Figure 3, there are a number of outliers shown by results not falling on the line of best fit.
   Seed mass was not controlled.
   Height of drop of seed was not controlled.
   Wing width/thickness was not controlled.
   The experimental data is not valid as some results were not controlled and so might have affected the results collected and/or because the data may contain errors.
   This means the students cannot be definite about their conclusion.
   However, there is a clear trend, so this could be confirmed with further work by the students/by looking at others' data.

Biology Chapter 3.6: Reach: Inheriting characteristics

Summary questions

1. Alleles, dominant, recessive, two [4 marks]
2. Freckles are most likely to be dominant [1 mark] as it is caused by a dominant allele, [1 mark] so only one copy of the allele is required. [1 mark]
3. Blue eyes are caused by a recessive allele, [1 mark] so two copies of the blue allele are required. [1 mark] Both parents must have one brown allele and one blue allele, [1 mark] and the child inherits the blue allele from the mum and the dad. [1 mark]

Biology Chapter 3.7: Reach: Asexual reproduction and cloning

Summary questions

1. Identical, clones, asexual, one [4 marks]
2. Asexual reproduction involves one parent whilst sexual reproduction requires two parents. [1 mark] There is no genetic variation in asexual reproduction whereas sexual reproduction involves the mixing of genetic material from the two parents. [1 mark] This means that there is variation between parents and different offspring in sexual reproduction whereas the offspring in asexual reproduction are identical to the parent. [1 mark]
3. Named example of cloning, e.g., plant cuttings. [1 mark] Advantages and disadvantages should be relevant to their example: [1 mark for each]
   Advantages:
   Offspring can be reproduced quickly.
   Cloning is cheap (for plants).
   Desired characteristics are replicated in the offspring.
   Cloning of a species can prevent extinction.
   Human cells can be cloned for disease research.
   Disadvantages:
   Ethical concerns.
   Reduction of gene pool.
   Lack of variation.
   This increases the risk of disease/environmental change leading to a population being killed off/this increases the likelihood of extinction.

Biology Chapter 3 Summary Answers

1. Two parents [1 mark], gametes [1 mark], fertilisation [1 mark], show variation [1 mark]
2. [Allow 2 marks if all four steps in the correct order, allow 1 mark if two steps followed correctly]
   Uterus lining is lost through the vagina
   Uterus lining starts to thicken
   Ovulation occurs
   If the female is not pregnant, uterus lining starts to break down.
3. Similarities: Both involve gametes/require two parents/ involve the mixing of genetic material/produce variation in the offspring/the ovaries produce the female gametes. [2 marks]
   Differences: In plants the male sex cell is pollen, in animals it is sperm/in plants the anther produces the male gamete, in animals it is the testes/in plants fertilisation occurs in the ovary, in animals it occurs in the fallopian tube/uterus. [2 marks]
4. **a)** Use a fan/hair dryer with variable speeds/move to different distances. [1 mark]
   b) (Horizontal) distance the seed travels. [1 mark]
   c) Any two from: height the seed is dropped from/mass of the seed/wing size/shape of wings. [2 marks]
   d) Likely to be invalid [1 mark] as a large spread suggests another variable has influenced the results. [1 mark]
5. Any six from [to get full marks students should include at least two named examples]:
   No form of contraceptive is 100% effective/all have a high degree of effectiveness.
   Condom has no side effects.
   Condom requires no medical approval.
   Condom protects against STIs.
   Pill may have side effects.
   Pill / contraceptives containing hormones have to be prescribed by a doctor.
   Pill has to be taken every day.
   Condom may reduce feelings of pleasure.

Accept any other relevant point/information about other named contraceptives.

6. a) Day 14 [1 mark]
   b) 2/3 days [1 mark]
   c) Between days 0–13/14, oestrogen levels rise [1 mark], reaching a peak level at day 13 /14. [1 mark] Oestrogen levels then fall. [1 mark]
   d) As oestrogen levels rise, the uterus lining builds up. [1 mark] This is needed for the egg to implant if fertilised. [1 mark] When oestrogen levels fall, it will trigger the lining of the uterus to breaks down. [1 mark]
7. a) A Only one parent is required [1 mark], D there is no variation in the offspring produced. [1 mark]
   b) Organism that is genetically identical to its parent. [1 mark]
8. [Allow 3 marks for all 4 correct, 2 marks for 2 or 3 correct, 1 mark for 1 correct]
   Gene – section of DNA that codes for a characteristic
   Allele – different versions of a gene
   Dominant allele – allele that will always be expressed if present
   Recessive allele – will be expressed if two copies of the allele are present

Biology Chapter 4: Variation and natural selection

Biology Chapter 4 Opener

Reactivate your knowledge questions

1. Photosynthesis; carbon dioxide and water (and light)
2. No living individuals of a species left anywhere in the world
3. A scar can be a characteristic affected by environmental, rather than inherited, variation.

Biology Chapter 4.2: Apply: Variation

Summary questions

1. Characteristics, inherited/genetic, inherit, environmental [4 marks]
2. Inherited - blood group, number of spots on a ladybird [1 mark] Environmental – scars [1 mark] Both - Fruit yield, freckles, leaf surface area [1 mark]
3. a) Inherited variation is caused by the genes/genetic material you inherit [1 mark], but environmental variation is a result of factors/conditions in your surroundings. [1 mark]
   b) Any two from: water availability, light availability, concentration of carbon dioxide, nutrient availability [2 marks]
4. Mass/height/skin tone/other acceptable answer [1 mark] For example, skin tone - natural skin colour is caused by your genes, [1 mark] sunlight in the environment can cause it to darken/tan. [1 mark]
5. a) 5.8 – 1.4 [1 mark] = 4.4 cm [1 mark]
   b) Change = 8.2 – 4.9, [1 mark] Percentage change = (8.2 – 4.9)/4.9 [1 mark] = 67% [1 mark]
6. Any six from:
   Height – mostly genetic/little effect due to environment
   Reasons: Little variation between identical twins (same genes) brought up in same environment
   and in different environments
   large difference between non-identical twins (different genes) in same environment
   IQ: mostly environment/little difference due to genes
   Reasons: Little variation between identical twins (same genes) in same environment
   and non-identical twins (different genes) in same environment.
   Large difference in identical twins (same genes) brought up in different environment.

Biology Chapter 4.3: Apply: More adaptations

Summary questions

1. Behavioural, functional, structural [3 marks]
2. Structural – large ears, bee markings [1 mark]
   Behavioural – making a nest, hunting in packs [1 mark]
   Functional – producing venom [1 mark]
3. Any three from: sharp teeth, sharp claws, good eyesight, good hearing, able to run/swim/fly rapidly, well camouflaged. [3 marks]
4. [Award 2 marks for named adaptations and 2 marks for linked explanations] For example, nocturnal – hunt for food when it is cooler, store water in body – allows animal to survive long periods without drinking.
5. a) Functional adaptation [1 mark]
   b) Any three from: uses sticky silk in parts of its web where it wants to capture insects/flies [1 mark] for food, [1 mark] uses non-sticky silk in places where it needs to walk across the web [1 mark] without getting itself stuck. [1 mark]
6. Any six from [to achieve maximum marks at least one adaptation must be stated from each category]:
   Structural:
   Long eye lashes – prevent sand entering the eyes
   Large feet – spready body mass/prevent sinking into the sand
   Camouflage – avoid being seen by predators
   Thick coat – cope with cold winter
   Behavioural:
   Search for food at night
   Urinate on legs to cool down
   Functional:
   Store fat – food reserves
   Very little sweating – avoid losing water
   Absorb food from water – maximise water up take.
   Allow any other relevant camel adaptations.

Biology Chapter 4.4: Apply: Investigating competition

Summary questions

1. Compete, sampling, quadrat, transect [4 marks]
2. a) Space/water [1 mark]
   b) Mates – as plants do not compete for mates/reproduce using the wind/insects, which carry the pollen between plants [1 mark] *or* Food – as plants make their own food by photosynthesis. [1 mark]
3. a) So they can photosynthesise/make their own food. [1 mark]
   b) Increase further from wall. [1 mark]
4. Any four from: [2 marks for four, 1 mark for two or three]
   Lay a tape measure in a straight line from the wall.
   Sample every X metres.
   Place a quadrat at the first sample site.
   Count the number of individual plants in the quadrat and which species they belong to.
   Repeat the previous two steps for the rest of the sampling site.
5. a) B [1 mark]
   b) Any three from:
   Nothing grows immediately next to the wall
   Plant A increases in abundance the further the distance from the wall
   Plant B increases in abundance up to 10m from the wall, then decreases again
   The further from the wall, the greater the overall coverage of plants
6. Any six from:

Method

1. Lay a tape measure in a straight line.
2. Ensure the tape measure crosses the footpath, and extends into the surrounding field.
3. Choose to take samples every X metres.
4. Place a quadrat at the start of the transect/on one side of the footpath.
5. Count the number of individual plants in the quadrat and which species they belong to.
6. Repeat steps 4 and 5 for the rest of the sampling site.
   Other marking points:
   Plot appropriate graphs/charts of the data.
   Compare data between the trodden and non-trodden areas.

Biology Chapter 4.5: Apply: Natural selection and antibiotic-resistant bacteria

Summary questions

1. Selection, species, characteristics/adaptations, resistance [4 marks]
2. Any three from: [3 marks]
   Wash hands thoroughly/use antiseptic gels
   Clean surfaces with disinfectants
   Use sterile medical equipment
   Use antiseptics to clean cuts
3. Clean with antiseptic [1 mark] to kill bacteria, [1 mark] cover with a sterile plaster [1 mark] to prevent microorganism entry. [1 mark]
4. a) Change in DNA. [1 mark]
   b) Most are harmful to the bacteria [1 mark] and cause the bacteria to die. [1 mark]
   As an alternative, allow that the DNA altered affects another characteristic
5. Any four from:
   Bacteria reproduce very rapidly/many generations can be produced in one day
   whereas human reproduction takes many years.
   Evolution occurs by the process of natural selection
   where those most adapted survive and reproduce/natural selection.
   This process requires many generations for the species to evolve.
6. Any six from:
   When a course of antibiotics is not completed, some bacteria are not killed in the patient
   and these multiply quickly.
   Bacterial reproduction increases the chance of mutations
   which occur when DNA is damaged/altered during replication.
   Most mutations result in the death of the bacterium
   but some mutations allow bacteria to become antibiotic-resistant.
   Eventually, the whole population of bacteria will become antibiotic-resistant.
   So your friend should get some replacement pills and finish the course of antibiotics.

Biology Chapter 4.6: Reach: Maintaining biodiversity

Summary questions

1. Biodiversity, banks, breeding, conservation [4 marks]
2. Advantages – populations of endangered species can be stabilized/increased [1 mark] and endangered species can then be re-introduced into their natural habitat. [1 mark]
   Disadvantages – it is difficult to maintain genetic diversity [1 mark] and organisms bred in captivity may not survive in the wild. [1 mark]
3. Example answer:
   Positive effects by humans: Populations of species can be maintained using a range of techniques, for example, conservation, captive breeding, and seed banks. Conservation is where humans protect natural habitats of endangered species. This reduces disruption to food chains and webs. Captive breeding can stabilize/increase the population of endangered species before re-introducing them to the wild. Seed banks provide a backup against the extinction of species by preserving their genetic material.
   Negative effects by humans: Human activity has caused many organisms to become extinct or endangered. For

example, hunting/poaching of animals in the case of the rhino. Deforestation leads to the loss of animal habitat. There is also general competition with animals (for food, space, and water).

Biology Chapter 4.7: Reach: Ecosystem – biotic and abiotic factors

Summary questions

1. Abiotic – rainfall, temperature
   Biotic – predators, competitors [Award 1 mark for each factor correctly sorted, max 4]
2. Any four from [allow two reasons and two linked explanations]:
   New species is a predator, [1 mark] eats the animals that currently live there. [1 mark]
   New species is a better competitor for food, [1 mark] so organisms of existing species die from lack of food. [1 mark]
   New species competes more successfully for space, [1 mark] so existing animals may have nowhere to live. [1 mark]
3. Examples include: [6 marks]
   Plants need light for photosynthesis/to make food.
   In high light intensities, photosynthesis rate is high, so plants grow well/quickly.
   Good food supply for animal species which feed on them, so animals survive and reproduce, increasing/maintaining population sizes.
   In low light intensities, photosynthesis rate is low, so plants do not grow well/do not survive.
   Food supply for animal species is reduced/removed, so animals do not have enough food to survive and decrease in number.

Biology Chapter 4 Summary Answers

1. [2 marks for three correct, 1 mark for one or two correct]

Structural adaptations	physical features that help survival and reproduction
Behavioural adaptations	responses made by an organism that help survival and reproduction
Functional adaptations	body processes that help survival and reproduction

2. **a)** Space, [1 mark] water. [1 mark]
   b) Plants make their own food by photosynthesis. [1 mark]
   c) Mates [1 mark]
3. Height/mass/number of fruits produced/other relevant example. [1 mark] For example, the typical yield of a particular species is determined by its genes, [1 mark] but receiving a sufficient amount of water/light in its environment will affect the yield. [1 mark]
4. Any four from:
   When bacteria multiply/replicate, their DNA may be damaged/altered.
   A change in DNA is called a mutation.
   The mutation could cause resistance to an antibiotic.
   When exposed to the antibiotic, those with the mutation survive and reproduce (passing the characteristic on).
   Those without the mutation are killed.
   Over time all the members of the population have the mutation/are antibiotic-resistant.
5. Any six from: [Allow a maximum of three marks from any one category]
   Structural
   White fur – camouflage
   Thick coat – insulation
   Waterproof coat – so they can enter water
   Large paws – spread weight on ice/snow.
   Behavioural
   Dormancy (accept hibernation) – to avoid worst weather/survive when prey not available
   Functional
   High fat diet – for energy/makes food available other animals do not consume.
6. Any six from:
   Take samples along a transect.
   Lay a tape measure in a straight line from the shore moving inland.
   Choose to take samples every X metres.
   Place a quadrat at the start of the transect/at each sample site along the tape measure.
   Count the number of individual plants in the quadrat and which species they belong to.
   Repeat this at each the sampling site.
   Plot appropriate graphs/charts of the data.
   Take measurements of other factors such as salinity of soil to compare with plants present.
7. [2 marks for three correct, 1 mark for one or two correct]

Conservation	Protect an organism's habitat
Seed bank	So new plants can be grown
Breeding programme	Increase animal numbers

8. Abiotic factors – A mineral availability [1 mark], E pH [1 mark]
   Biotic factors – B bacteria present [1 mark], D competitors [1 mark]

Chemistry Chapter 1: Particle model and state change

Chemistry Chapter 1 Opener

Reactivate your knowledge questions

1. 

2. The sugar molecules spread out to form a solution
3. Melting

Chemistry Chapter 1.2: Apply: The particle model

Summary questions

1. Motion, random, gas, particles, solution [5 marks]
2. a) Movement – random, sliding over each other, [1 mark] separation – touching neighbours. [1 mark]
   b) Particles have broken off the solid and spread apart in the water to make a solution. [1 mark]
   c) Might be poisonous/toxic/cause harm. [1 mark]
3. a) To try to ensure that the results were reliable, and not a one-off occurrence. [1 mark]
   b) To eliminate the possibility that the pollen was moving itself as a result of its being alive. [1 mark]
   c) Einstein [1 mark]
4. a) The glitter would start moving around constantly and randomly/Brownian motion. [1 mark]
   b) The salt would dissolve in the water. [1 mark]

Chemistry Chapter 1.3: Apply: Substances

Summary questions

1. One type, the same, compounds, nitrogen, carbon dioxide [5 marks]
2. a) It is an element/it has one type of particle. [1 mark]
   b) It is a mixture of two substances/elements. [1 mark]
3. a) x-axis correctly labelled 'country', [1 mark] y-axis correctly labelled 'mass of sugar in 330 cm^3 of fizzing orange, in g', [1 mark] even scale on y-axis, [1 mark] all bars the correct height. [2 marks – award 1 mark if at least half the bars are correct heights]
   b) UK [1 mark]
   c) Pakistan and India [1 mark – both countries required for the mark]
   d) For example – in some countries, people like sweeter drinks than in others/some countries have stricter laws about sugar content in food than others. [1 mark]

Chemistry Chapter 1.4: Apply: States of matter

Summary questions

1. Gas, solid, quickly/fast, touch, liquid [5 marks]
2. a) Vibrating on the spot [1 mark] in regular pattern. [1 mark]
   b) Moving quickly and randomly from place to place, [1 mark] randomly arranged and far apart. [1 mark]
3. Motion: in both states the particles are moving from place to place [1 mark], but in the gas state the particles move throughout the pocket and in the liquid state the particles slide over each other. [1 mark]
   Arrangement: in both states, the particles are randomly arranged [1 mark], but in the gas state the particles are far apart and in the liquid state the particles touch their neighbours. [1 mark]
4. a) J, D, C [1 mark]
   b) If the people who breathe out the most droplets also breathe out the most virus, these people are likely to be more infectious/infect a greater number of people. [1 mark]

Chemistry Chapter 1.5: Apply: Sublimation: a change of state

Summary questions

1. Solid, gas, liquid, deposition [4 marks]
2. Movement – changes from vibrating on the spot in the solid [1 mark], to moving randomly and fast in all directions in the gas. [1 mark] Arrangement – changes from a regular pattern with particles touching [1 mark] to a random arrangement with particles far apart. [1 mark]
3. In both changes, the substance ends up in the gas state [1 mark] and the particles leave the surface. [1 mark] In evaporation, the starting state is liquid, but in sublimation the starting state is solid. [1 mark]
4. a) Dry ice can keep medicines cold. [1 mark]
   b) Dry ice stays cooler for longer than ice, [1 mark] melted ice could damage medicines. [1 mark]
5. a) 68% [2 marks]
   b) Percentage remaining = 87% [1 mark]
   87% × 200 g = 174 g [2 marks]

Chemistry Chapter 1.6: Reach: Energy transfer in changes of state

Summary questions

1. From, melting, evaporating, to, freezing, condensing [6 marks]
2. Energy must be transferred from the surroundings to the solid [1 mark] so that its particles can overcome the strong forces between them [1 mark] and escape to form a gas. [1 mark]
3. a) Aluminium [1 mark]
   b) Aluminium has the higher melting point, [1 mark] so more energy is required to overcome the forces between the particles in aluminium. [1 mark]

Chemistry Chapter 1.7: Reach: Limitations of the particle model

Summary questions

1. An idea that explains observations and helps in making predictions. [1 mark]
2. You can pour a substance when it is liquid because its particles move all the time, sliding over each other. [1 mark] In the solid state, the particles are in fixed positions. [1 mark]
3. Particles are not all spheres [1 mark] and particles are not solid. [1 mark]
4. It does not explain all observations perfectly. [1 mark] Predictions made with the model are not always accurate. [1 mark]

Chemistry Chapter 1 Summary Answers

1. Gas [1 mark], melting [1 mark], liquid [1 mark]
2. **a)** A material that is not a mixture. It has the same properties all the way through. [1 mark]
   b) Oxygen, [1 mark] sodium chloride [1 mark], carbon dioxide. [1 mark]
3. **a)** Vibrating on the spot. [1 mark]
   b) Particles move randomly from place to place, sliding over other particles. [1 mark] Particles touch their neighbours and are not in a regular pattern. [1 mark]
4. **a)** Particles move quickly and randomly everywhere in the pocket of gas. [1 mark] Particles are far apart from each other and not in a regular pattern. [1 mark]
   b) **i)** Between 0.85 and 0.9. [1 mark]
   ii) Between 220 Hz and 240 Hz. [1 mark]
   iii) The graph shows that the absorption of the 50 mm foam is greater than the absorption of the 12 mm foam, until 2000 Hz, where the lines cross. [1 mark]
   iv) There is more material, including more air pockets, to absorb the sound waves. [1 mark]
5. **a)** To ensure that the pollen was not moving because it was alive. [1 mark]
   b) To see if the same result occurred with particles that had never lived. [1 mark]
6. **a)** Air particles collide randomly with the pieces of dust, making them move. [1 mark]
   b) Brownian motion. [1 mark]
7. Freezing, [1 mark] condensing [1 mark]
8. Model – particles are spherical, reality – particles can be different shapes. [1 mark]
   Model – particles are solid, reality – particles are mainly empty space. [1 mark]

Chemistry Chapter 2: Atoms and the Periodic Table

Chemistry Chapter 2 Opener

Reactivate your knowledge questions

1. Water contains more than one element, hydrogen and oxygen, so cannot be an element.
2. Gold is unreactive, shiny, and conducts electricity; a single gold atom does not have these properties.
3. C, Ca, Cs, Cl, Cu, Fe

Chemistry Chapter 2.2: Apply: The development of the Periodic Table

Summary questions

1. 60, mass, a conference, Mendeleev [4 marks]
2. **a)** Gallium, scandium, germanium, rhenium [4 marks]
   b) Scandium – Lars Nilson [1 mark] germanium – Clemens Winkler [1 mark] rhenium – Ida Tacke and Walter Noddack [1 mark]
   c) The discovery of missing elements showed that Mendeleev's Periodic Table had already made successful predictions. [1 mark] This indicates that it was a successful theory and could correctly make other predictions. [1 mark]
3. For example – some elements are invisible; [1 mark] some elements are strongly joined to other elements in compounds; [1 mark] many experimental techniques were not available to scientists at that time. [1 mark]

Chemistry Chapter 2.3: Apply: Inside atoms

Summary questions

1. Nucleus, protons, neutrons, electrons [4 marks]
2. **a)** Nucleus [1 mark]
   b) 3 (number of electrons = number of protons) [1 mark]
   c) 3 + 4 = 7 [2 marks, 1 mark for correct working but incorrect answer]
3. **a)** The football represents a nucleus in the gold foil [1 mark] and the ping-pong balls represent positively charged particles being fired at the gold foil. [1 mark]
   b) For example: Similarity – only a few ping-pong balls will hit the football and bounce straight back (just as only a few positively charged particles hit a nucleus and bounce back). [1 mark] Difference – the football and the ping-pong balls have no electrical charge, [1 mark] but in fact the nucleus is positively charged and the particles being fired at it are also positively charged. [1 mark]

Chemistry Chapter 2.4: Apply: Metals and non-metals in the Periodic Table

Summary questions

1. Groups, metal, 1, left, non-metal, 0, right [7 marks]
2. For example – good conductor of thermal energy [1 mark], good conductor of electricity [1 mark], shiny when freshly cut [1 mark], solid at room temperature. [1 mark]
3. Non-metal [1 mark], because its properties are typical of non-metal elements. [1 mark]

Chemistry Chapter 2.5: Apply: Groups of the Periodic Table

Summary questions

1. 2, decreases, decreases, hydrogen [4 marks]
2. **a)** Names of metals on x-axis, [1 mark] with axis labelled 'Metal' or 'Element'. [1 mark]
   Evenly spaced scale on y-axis, [1 mark] with axis labelled 'Hardness, Moh scale'. [1 mark]

Correctly labelled y-axis, with units [1 mark] and bars correctly plotted. [1 mark]

b) Hardness decreases from top to bottom of the group. [1 mark]

3. a) Magnesium chloride, [1 mark] hydrogen. [1 mark]

b) magnesium + hydrochloric acid → magnesium chloride + hydrogen [3 marks]

4. a) For example, three test tubes half-filled with water. [1 mark]

b) Add one small piece of a different Group 1 element to each test tube. [1 mark]
To make the experiment fair, use pieces of the same size/same amount of water/ water of the same temperature. [1 mark]

c) The metal which makes the most vigorous bubbling is the most reactive metal. [1 mark]

Chemistry Chapter 2.6: Apply: Elements, compounds, and mixtures

Summary questions

1. The same, two, strongly, mixture, are important [5 marks]

2. a) 1 : 3 [1 mark]

b) 1 : 4 [1 mark]

3. a) TiC [2 marks]

b) TiS_2 [2 marks]

c) $BaTiO_3$ [3 marks]

4. Gold is unreactive [1 mark] and shiny. [1 mark]

Chemistry Chapter 2.7: Reach: Electronic structure

Summary questions

1. Negative, shells, chemical [3 marks]

2. a) Fluorine – 2 electrons in inner shell, and 7 in outer shell [1 mark]
chlorine – 2 electrons in inner shell, 8 in next shell, and 7 in outer shell. [1 mark]

b) Chlorine – 7 [1 mark]
fluorine – 7 [1 mark]

c) Their atoms have the same number of electrons in their outer shells. [1 mark]

3. a) 2 [1 mark]

b) Silicon – 2 electrons in inner shell, 8 in next shell, 4 in outer shell [1 mark]
sulfur – 2 electrons in inner shell, 8 in next shell, 6 in outer shell. [1 mark]

c) Different chemical properties [1 mark] because the atoms of each element have different numbers of electrons in their outer shells/silicon has 4 electrons in its outer shell, but sulfur has 6 electrons in its outer shell. [1 mark]

Chemistry Chapter 2.8: Reach: Bonding

Summary questions

1. Covalent, ions, ionic [3 marks]

2. a) Bonds formed when atoms share electron pairs. [1 mark]

b) Bonds formed between oppositely charged ions as a result of electrostatic forces that act in all directions. [1 mark]

3. The outer shell electron from a sodium atom is transferred to the outer shell of a chlorine atom. [1 mark] The sodium atom now has one more positively charged protons than negative electrons, so has a single positive charge overall. [1 mark] The chlorine atom now has one more negatively charged electrons than positively charged protons, so has a single negative charge overall. [1 mark] Diagram of positive sodium ion and negative chlorine ion. [1 mark]

4. a) Covalent [1 mark] because oxygen is a non-metal. [1 mark]

b) Ionic [1 mark] because compound is made up of atoms/ions of a metal and non-metal. [1 mark]

c) Covalent [1 mark] because compound is made up of atoms of two non-metal elements. [1 mark]

Chemistry Chapter 2 Summary Answers

1. Neutrons, +1, electrons, -1, protons [5 marks]

2. C – not shiny in the solid state, [1 mark] D – poor conductors of thermal energy. [1 mark]

3. a) Bar chart – y-axis labelled 'melting point in °C'; [1 mark] regular scale on y-axis, from 0 to 1300; [1 mark] bars correctly drawn. [1 mark]

b) As you go down Group 2, the melting points of the elements decrease. [1 mark] Magnesium does not fit the pattern. [1 mark]

c) Between 700 °C and 750 °C) [1 mark]

4. a) C [1 mark]

b) A and D [1 mark] because they are not shiny, do not conduct electricity and have low melting points. [1 mark]

c) B and C [1 mark]

d) D [1 mark]

5. NH_3, [1 mark] Al_2O_3, [1 mark], H_2O, [1 mark] CO_2, [1 mark] NO [1 mark]

6. a) Independent variable – element; [1 mark] dependent variable – how vigorously the element reacts with water. [1 mark]

b) Any one from: volume of water, mass of element, surface area of element, temperature. [1 mark]

c) Because the reactions would be very vigorous and hazardous. [1 mark]

7. a) Covalent [1 mark]

b) Electrostatic [1 mark]

8. a) Two concentric circles, with 2 crosses/dots on the inner circle and 7 on the outer circle. [1 mark]

b) 7 [1 mark]

Chemistry Chapter 3: Chemical changes

Chemistry Chapter 3 Opener

Reactivate your knowledge questions

1. 44 g
2. Reacts to make
3. Energy is transferred from the surroundings

Chemistry Chapter 3.2: Apply: Chemical and physical changes

Summary questions

1. Chemical, do not change, does [3 marks]
2. **a)** Before – liquid, [1 mark] after – gas/vapour [1 mark]
   b) 5 [1 mark]
   c) Arrangement – from touching neighbours [1 mark] to far apart [1 mark]; movement – from sliding over each other [1 mark] to moving randomly in all directions throughout the container [1 mark]
3. **a)** 7 [1 mark]
   b) The oxygen atom has come from the carbonate particle [1 mark]; the hydrogen atoms have come from the fumarate molecule. [1 mark]
4. In a chemical reaction, the particles change [1 mark] because the atoms rearrange and join together differently. [1 mark] In a physical change, the particles do not change [1 mark], but their movement and arrangement change. [1 mark]

Chemistry Chapter 3.3: Apply: Reactants and products

Summary questions

1. Reactants, products, water [3 marks]
2. **a)** Carbon, [1 mark] hydrogen. [1 mark]
   b) Carbon dioxide and water [1 mark]
   c) Combustion [1 mark]
3. methane + oxygen → carbon dioxide + water [3 marks]
4. Carbon dioxide [1 mark] and water [1 mark]

Chemistry Chapter 3.4: Apply: Writing chemical equations

Summary questions

1. Right, formulae, ratio [3 marks]
2. **a)** Carbon dioxide – 3, [1 mark] water – 4 [1 mark]
   b) 3 : 4 [1 mark]
   c) 1 : 3 [1 mark]
   d) One propane molecule, as shown in Figure 4; [1 mark] 5 oxygen molecules, each with two red atoms; [1 mark] 3 carbon dioxide molecules, each with one black carbon atom joined to two red oxygen atoms; [1 mark] and 4 water molecules, each with one red oxygen atom joined to two white hydrogen atoms. [1 mark]
3. **a)** $N_2 + 2O_2 \rightarrow 2NO_2$ [4 marks]
   b) $C_5H_{12} + 8O_2 \rightarrow 5CO_2 + 6H_2O$ [4 marks]

Chemistry Chapter 3.5: Apply: Conservation of mass

Summary questions

1. Products, mass, same [3 marks]
2. Figure 3 shows 20 tablets, each of 0.5 g, so total mass is 0.50 g × 20 = 10 g [2 marks]
3. Aminophenol 0.72 g × 1000 = 720 g; ethanoic anhydride 0.68 g × 1000 = 680 g [3 marks]
4. **a)** Mass of reactants = 44 g + 160 g = 204 g [1 mark]
   Mass of reactants = mass of products
   So mass of water = 204 g – 132 g = 72 g [1 mark]
   b) 160 g × 2 = 320 g [2 marks]
   c) 132 g ÷ 2 = 66 g [2 marks]

Chemistry Chapter 3.6: Apply: Energy in chemical reactions

Summary questions

1. Energy, to [2 marks]
2. So that the test is fair. [1 mark]
3. 63 °C – 22 °C = 41 °C [2 marks, 1 for working and 1 for answer]
4. **a)** Bar chart with names of food on the x-axis [1 mark] and temperature change on the y-axis, with even scale. [1 mark] Bars correctly drawn. [1 mark]

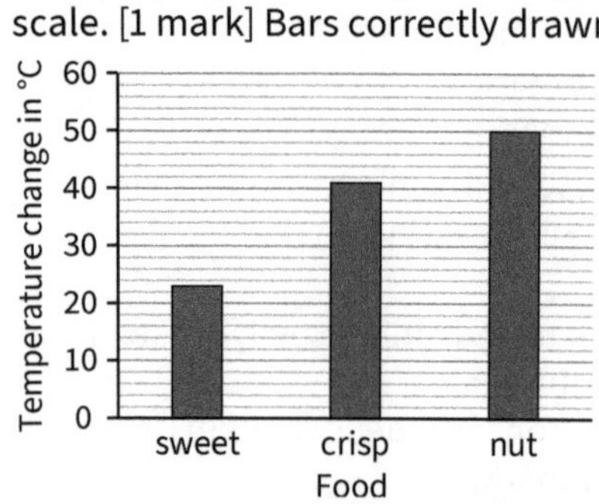

   b) The burning sweet heated up the water the least, [1 mark] so it transferred less energy to the water than the crisp or nut. [1 mark]
   c) Burning 1 g of nut made the water hotter than burning 1 g of crisp or 1 g of burning jelly sweet. [1 mark] This means that the nut transferred most energy to the water [1 mark] and the jelly sweet transferred least energy. [1 mark]

Chemistry Chapter 3.7: Reach: Exothermic and endothermic

Summary questions

1. To, increases, exothermic, exothermic [4 marks]
2. a) [2 marks]

Temperature before reaction, in °C	19
Temperature after reaction, in °C	36
Temperature change, in °C	17

   b) Exothermic [1 mark] because during the reaction, the energy from the reaction heats up the reaction mixture. [1 mark]
3. a) Endothermic [1 mark] because during the reaction, the reacting mixtures take energy from the solution. [1 mark]
   b) Take equal amounts/masses of the two substances [1 mark] and dissolve each in an equal volume of water, each starting at the same temperature. [1 mark] Measure the temperature change when the substance dissolves. [1 mark] The substance that dissolves with the greater temperature change transfers more energy. [1 mark]

Chemistry Chapter 3 Summary Answers

1. Chemical reaction, reactants, products, combustion [4 marks]
2. a) Reactants – sodium and iodine [1 mark] product – sodium iodide [1 mark]
   b) Reactants – carbon and oxygen [1 mark] product – carbon monoxide [1 mark]
   c) Reactants – zinc and hydrochloric acid [1 mark] products – zinc chloride and hydrogen [1 mark]
   d) Reactants – lithium and water [1 mark] products – lithium hydroxide and hydrogen [1 mark]
3. a) Reactants – hydrogen and chlorine [1 mark] product – hydrogen chloride [1 mark]
   b) Green [1 mark]
   c) 2 [1 mark]
   d) 2 [1 mark]
   e) Because the number of hydrogen and chlorine atoms are same [1 mark] in the reactants and in the product. [1 mark]
4. a) zinc + oxygen → zinc oxide [3 marks]
   b) calcium carbonate → calcium oxide + carbon dioxide [3 marks]
   c) sodium bromide + silver nitrate → silver bromide + sodium nitrate [3 marks]
5. a) 4 g + 32 g = 36 g [2 marks]
   b) 1.2 g + 3.2 g = 4.4 g [2 marks]
   c) 88 g – 56 g = 32 g [2 marks]
   d) 64 g – 32 g = 32 g [2 marks]
6. 10 CO_2 molecules, each with one black circle in the middle and two red circles joined on either side. [2 marks] 11H_2O molecules, each with one red circle in the middle and two white circles joined to it. [2 marks]
7. a) $2Mg + O_2 \rightarrow 2MgO$ [1 mark]
   b) $4Na + O_2 \rightarrow 2Na_2O$ [1 mark]
   c) $Zn + 2HCl \rightarrow ZnCl_2 + H_2$ [1 mark]
   d) $CH_4 + 2O_2 \rightarrow CO_2 + 2H_2O$
   e) $Fe_2O_3 + 3CO \rightarrow 2Fe + 3CO_2$
8. To, exothermic, endothermically [3 marks]

Chemistry Chapter 4: Useful chemical reactions

Chemistry Chapter 4 Opener

Reactivate your knowledge questions

1. For example, calcium reacts with cold water, zinc burns brightly with air.
2. For example, iron is attracted to magnets, copper is not.
3. Mercury

Chemistry Chapter 4.2: Apply: Using metals

Summary questions

1. Physical and chemical, does not, chemical [3 marks]
2. a) chromium or fluorine [1 mark]
   b) chromium fluoride [1 mark]
   c) Reacts to make. [1 mark]
3. a) chromium oxide [1 mark]
   b) chromium chloride and hydrogen [1 mark]
4. a) chromium + bromine → chromium bromide [3 marks]
   b) chromium + sulfuric acid → chromium sulfate + hydrogen [3 marks]
   c) chromium + chlorine → chromium chloride [3 marks]
5. Similarities – both are shiny grey [1 mark] differences – chromium is harder than palladium, [1 mark] palladium has a lower melting point than chromium [1 mark]
6. Physical – shiny grey appearance [1 mark] chemical – does not react with oxygen from the air at room temperature. [1 mark]

Chemistry Chapter 4.3: Apply: The reactivity series

Summary questions

1. Reactivity, reactive, unreactive, independent, dependent [5 marks]
2. a) A – Hypothesis includes a prediction and a scientific reason [1 mark]
   b) The order of reactivity of the metals tested decreases from zinc, to iron, to lead, to copper. [1 mark] The results show this because the reactions of the metals with hydrochloric acid get less vigorous from zinc to copper. [1 mark]
   c) No reaction because copper did not react with hydrochloric acid [1 mark] and palladium is below copper in the reactivity series. [1 mark]
3. Zinc [1 mark]

Chemistry Chapter 4.4: Apply: Displacement reactions

Summary questions

1. Less, compound, more [3 marks]
2. A, C, E [3 marks]
3. magnesium + iron sulfate → magnesium sulfate + iron [3 marks] calcium + magnesium oxide → calcium oxide + magnesium [3 marks] lead + copper oxide → lead oxide + copper [3 marks]
4. iron, X, silver [2 marks]

Chemistry Chapter 4.5: Apply: Extracting metals

Summary questions

1. Below, displacement, electrolysis [3 marks]
2. a) Magnesium is more reactive than carbon, so carbon cannot displace magnesium from its compounds. [1 mark] Electrolysis is therefore chosen, even though it is expensive. [1 mark]
   b) Iron is less reactive than carbon, so carbon can displace iron from its compounds. [1 mark] There is no point in using expensive electrolysis if an element can be extracted by heating with cheap carbon. [1 mark]
3. a) Electrolysis, because potassium is more reactive than carbon, [1 mark] so cannot be extracted by heating with carbon. [1 mark]
   b) Heating with carbon, because zinc is less reactive than carbon, so the method is chemically possible. [1 mark] Electrolysis could also be used, but is not chosen because it is more expensive. [1 mark]
   c) Heating with carbon, because nickel is less reactive than carbon, so the method is chemically possible. [1 mark] Electrolysis could also be used, but is not chosen because it is more expensive. [1 mark]

Chemistry Chapter 4.6: Apply: Using metals: catalysts

Summary questions

1. Nitrogen, catalytic converter, platinum/palladium/rhodium, catalysts, less [5 marks]
2. a) Platinum [1 mark], palladium [1 mark], rhodium [1 mark]
   b) As a catalyst/to speed up chemical reactions. [1 mark]
   c) carbon dioxide [1 mark] and nitrogen. [1 mark]
   d) On the surface. [1 mark]
   e) For example, carbon monoxide is poisonous [1 mark], so causes greater immediate harm that carbon dioxide. [1 mark]
3. Titanium, tungsten, and vanadium [1 mark] because they are transition metals and many transition metals are good catalysts. [1 mark]

Chemistry Chapter 4.7: Apply: Relative mass

Summary questions

1. Neutrons, 7, 28 [3 marks]
2. a) $14 \times 2 = 28$ [2 marks]
   b) $19 \times 2 = 38$ [2 marks]
   c) $14 + 16 = 30$ [2 marks]
   d) $(2 \times 12) + (6 \times 1) = 30$ [2 marks]
3. a) 2 kg = 2000 g [1 mark]
   Mass in one day = $(2000 \times 44) \div 16$ [1 mark] = 5500 g (5.5 kg) [1 mark]
   b) 5.5×365 [1 mark] = 2008 kg [1 mark]

Chemistry Chapter 4.8: Apply: Percentage yield

Summary questions

1. Theoretical, actual, divided, multiplied [4 marks]
2. a) $\frac{2.0\text{ g}}{2.5\text{ g}} \times 100\% = 80\%$ [2 marks]
   b) $\frac{6.4\text{ g}}{8.0\text{ g}} \times 100\% = 80\%$ [2 marks]
   c) $\frac{96\text{ g}}{108\text{ g}} \times 100\% = 88.9\%$ [2 marks]
3. $180 \times 2 = 360$ g [2 marks]

Chemistry Chapter 4 Summary Answers

1. a) Sonorous. [1 mark]
   b) A good conductor of thermal energy. [1 mark]
   c) A good conductor of electricity. [1 mark]
   d) Strong. [1 mark]
2. a) To convert harmful exhaust gases to less harmful ones. [1 mark]
   b) Any two of: platinum, palladium, rhodium. [2 marks]
3. a) It melts at 1063 °C. [1 mark]
   b) It is a good conductor of electricity [1 mark] It is a good conductor of thermal energy. [1 mark]
   c) It is always shiny. [1 mark]
   d) It melts at 1063 °C. [1 mark]
   It is a good conductor of electricity. [1 mark]
4. a) sodium [1 mark]
   b) Any one of: iron, zinc, magnesium, or sodium. [1 mark]
   c) Nothing/no reaction [1 mark]
   d) sodium and magnesium [1 mark]
5. a) A, B and E [1 mark]
   b) $Mg + CuSO_4 \rightarrow MgSO_4 + Cu$ [3 marks]
   $Zn + CuCl_2 \rightarrow ZnCl_2 + Cu$ [3 marks]
   $2Mg + PbO_2 \rightarrow 2MgO + Pb$ [3 marks]
6. a) $2 \times 19 = 38$ [1 mark]
   b) $1 + 19 = 20$ [1 mark]
   c) $12 + (2 \times 16) = 44$ [1 mark]

d) 14 + 16 = 30 [1 mark]
  a) 14 + (2 × 16) = 46 [1 mark]
  b) (2 × 14) + 16 = 44 [1 mark]

7. $\frac{1.8}{2.0} \times 100 = 90\%$ [2 marks]

Physics Chapter 1: Forces and motion

Physics Chapter 1 Opener

Reactivate your knowledge questions

1. For example, if you push a book along a desk.
2. The ground/your chair
3. Mass = kilogram, weight = newton

Physics Chapter 1.2: Apply: Forces and interactions

Summary questions

1. Third, pairs, interactions, size, different, Earth [6 marks]
2. a) For every action there is an equal and opposite reaction/forces come in interaction pairs. [1 mark]
   b) Interactions with the table, [1 mark] Earth, [1 mark] and string. [1 mark]
   c) Dot with arrow down labelled 'force of the Earth on the block'. [1 mark] Arrow upwards labelled 'force of the string on the block'. [1 mark] Arrow upwards labelled 'force of the table on the block'. [1 mark]
3. a) The force of the Earth on you and the force of you on the Earth. [1 mark]
   b) When you push down on the Earth with your foot, there is the force of you on the ground and the force of the ground on you [1 mark], which is bigger than the force of the Earth on you, so you move upwards. [1 mark]
4. When you blow up a balloon and let go, there is the force of the balloon on the air and the force of the air on the balloon. [1 mark] The force of the air on the balloon moves the balloon forwards. [1 mark] This is the same as the force of the air on the rocket pushing the rocket forward, [1 mark] which is part of an interaction pair with the force of the rocket on the air. [1 mark]
5. When you walk you are interacting with the ground [1 mark] since you push down on the ground and the ground pushes up on you. [1 mark] There is friction between you and the ground [1 mark] and if the friction is large enough, then the force on the ground will produce a force on you. [1 mark] This means you will move forward. [1 mark] If there is ice there is not enough friction to produce a forward force. [1 mark]

Physics Chapter 1.3: Apply: Mass, weight, and fields

Summary questions

1. Mass, kilograms, gravitational, weight, weight, mass [6 marks]
2. a) If the gravitational field is stronger, the lines are closer together. [1 mark]
   b) The gravitational field strength decreases. [1 mark]
   c) The distance from the Earth's centre (of mass) to the Earth's surface is roughly constant across the Earth's surface, [1 mark] so the gravitational field strength is assumed to be constant. [1 mark]
3. g = weight/mass
   = 300 N/50 kg [1 mark]
   = 6 N/kg [1 mark answer, 1 mark unit]
4. Each measurement would be bigger by 0.5 newtons. [1 mark] When you plot it, that means every point would be above where it is by 0.5 newtons. [1 mark] The line would be parallel with the line on the graph [1 mark] with a y-intercept of 0.5 newtons. [1 mark]
5. a) i. kilograms (kg)
      ii. newtons (N)
   b) Mass is a measure of the amount of stuff or material [1 mark] while weight is a measure of the force of the Earth or another large object. [1 mark] Mass does not change [1 mark] while the weight depends on the gravitational field that you are in. [1 mark]

Physics Chapter 1.4: Apply: Speed and distance–time graphs

Summary questions

1. Total distance, total time, metre, second, running, cycling, 8 [7 marks]
2. a) The slopes do not change. [1 mark]
   b) The slope of the graph for the fastest animal will be steepest. [1 mark]
3. Speed = distance/time = 10 m/2 s [1 mark]
   = 5 m/s [1 mark answer, 1 mark unit]
4. Graph will be a downward half-parabola [1 mark] starting from the origin with a steep gradient (high upward speed). [1 mark] The gradient gets less steep as the distance increases and (ball slows down) [1 mark] until the curve is horizontal at the ball's highest point (ball stops). [1 mark]
5. a) The line goes back to zero after a certain time [1 mark] so the object has returned to where it started from. [1 mark]
   b) Positive = walking away from home [1 mark] negative = walking towards home [1 mark]
      speed = distance/time = 8 m/4 s [1 mark]
      = 2 m/s [1 mark answer, 1 mark unit]
      speed = distance/time = 8 m/3 s [1 mark]
      = 2.7 m/s [1 mark answer, 1 mark unit]

6. Using tangents to the curve at 2 seconds and 8 seconds
   At 2 seconds, speed = distance/time = 2 m/4 s [1 mark]
   = 0.5 m/s [1 mark answer, 1 mark unit]
   At 8 seconds, speed = 10 m/5 s [1 mark]
   = 2m/s [1 mark answer, 1 mark unit]
   Change in speed = 2 m/s – 0.5 m/s [1 mark]
   = 1.5 m/s [1 mark]

Physics Chapter 1.5: Apply: Balanced and unbalanced

Summary questions

1. Cancel, balanced, stationary/still, steady speed, unbalanced, direction [6 marks]
2. **a)** No, [1 mark] her speed changes. [1 mark]
   b) No, [1 mark] her speed changes. [1 mark]
   c) Yes, [1 mark] her speed is not changing. [1 mark]
   d) Yes, [1 mark] her speed is not changing. [1 mark]
3. **a)** The direction of motion is constantly changing. [1 mark]
   b) The normal force of their seat/the tension force in a seatbelt. [1 mark]
4. 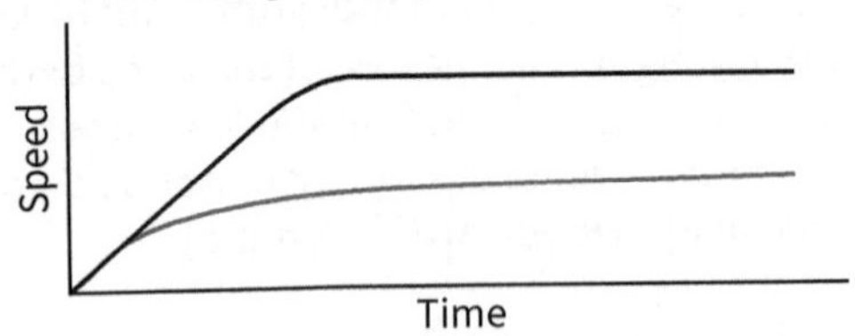

   Light grey line = thicker liquid, dark grey line line = thinner liquid
   a) Long acceleration [1 mark] that flattens off. [1 mark]
   b) Smaller terminal velocity [1 mark] that is reached quicker. [1 mark]
   c) With the thicker liquid, the plasticene will reach terminal velocity faster and the velocity will be smaller. [1 mark]
5. An object with a large inertia has a large mass. [1 mark] When you are pushing a car, [1 mark] it requires a large force to make the object move, [1 mark] so you would notice the large inertia of the car. [1 mark]
6. Mass is a measure of the amount of stuff or material and is measured in kilograms. [1 mark] Weight is a measure of the force of the Earth or another large object and is measured in newtons. [1 mark] Mass does not change [1 mark] while the weight depends on the gravitational field that you are in. [1 mark] For example, an astronaut would have a higher weight on Earth than on the Moon, [1 mark] but their mass would stay the same. [1 mark]

Physics Chapter 1.6: Reach: Resultant forces

Summary questions

1. [All correct 4 marks, 3 correct – 3 marks, 2 correct – 2 marks, 1 correct – 1 mark]
   vector – has magnitude and direction
   scalar – has magnitude only
   resultant force – the single force that replaces all the forces
   example of vector – displacement
   example of scalar – mass
2. **a)** Gravity/backpack's weight produces the force downwards, [1 mark] air resistance produces the force upwards. [1 mark]
   b) Resultant force = +40 N + (–5 N) [1 mark]
   = +35 N [1 mark answer, 1 mark sign]
   c) Resultant force = +10 N + (–20 N) [1 mark]
   = –10 N [1 mark answer, 1 mark sign]
3. The mass of a liquid has magnitude and no direction [1 mark] whereas the weight of a liquid has magnitude and direction (towards the Earth). [1 mark]
4. Your speed is not changing but your direction is, [1 mark] so your velocity is changing. [1 mark] The force is produced by the friction between the tyres and the road [1 mark] which continuously changes the velocity. [1 mark]

Physics Chapter 1.7: Reach: Acceleration and speed–time graphs

Summary questions

1. Speed, time, metres per second squared, large/big, zero [5 marks]
2. **a)** The acceleration in the first 10 seconds is changing [1 mark] and getting smaller. [1 mark]
   The acceleration in the last 10 seconds is constant. [1 mark]
   b) = change in speed/ time to change
   = (0 m/s – 20 m/s)/10s [1 mark]
   = –2 m/s^2 [1 mark sign, 1 mark answer, 1 mark unit]
3. Draw a tangent to the curve. [1 mark]
   a) = change in speed/time to change
   = (25 m/s – 0 m/s)/10s [1 mark]
   = +2.5 m/s^2 [1 mark sign, 1 mark answer, 1 mark unit]

Physics Chapter 1 Summary Answers

1. Mass, charge, magnetic, equilibrium [4 marks]
2. A – Forces arise due to interactions with objects, [1 mark]
   D– Forces can be shown by arrows [1 mark]
3. **a)** Average speed = total distance/total time [1 mark]
   = 800 m/180 s [1 mark]
   = 4.4 m/s [1 mark]
   b) Moving fastest at the steepest section, so D. [1 mark]
   speed = distance/time
   = 800 m – 600 m/(180 s-170 s) [1 mark]
   = 20 m/s [1 mark]
   c) A car. [1 mark]
4. **a)** Density of liquid. [1 mark]
   b) Time to fall. [1 mark]
   c) Two from: length of cylinder, type of ball, temperature of liquid. [2 marks]

d) As the density of the liquid increases, the time increases [1 mark] since there are more particles/more collisions with the ball to slow it down. [1 mark]

e) It is the same each time, so it is a systematic error. [1 mark]

5. a) g = W/m [1 mark]
= 15 N/0.4 kg = 37.5 N/kg (or 38)[1 mark value, 1 mark unit]

b) Mass = 0.4 kg [1 mark]

c) No, weight refers to the force of a large object on a small object. [1 mark]

6. a) Gravity/gravitational force. [1 mark]

b) Smaller [1 mark] since the gravitational field is weaker/gets smaller as you move away from the Earth. [1 mark]

c) The floor is not pushing up on them. [1 mark]

7. a) The force of the Earth on the leaf/leaf on the Earth [1 mark] and the force of the tree on the leaf/leaf on the tree. [1 mark]

b) Zero, [1 mark] it is not moving. [1 mark]

c) 0.1 N, [1 mark] the only force acting is the force of the Erath on the leaf, which is its weight. [1 mark]

d) Zero, [1 mark] its speed is not changing, so there is no resultant force. [1 mark]

Physics Chapter 2: Energy

Physics Chapter 2 Opener

Reactivate your knowledge questions

1. For example, gravitational potential, kinetic, chemical
2. Conduction, convection, radiation, force
3. 150 N

Physics Chapter 2.2: Apply: Energy stores

Summary questions

1. [5 marks, 1 mark per correct answer]
fuels – chemical
moving objects – kinetic
raised objects – GPE
hot objects – thermal
the Sun – chemical

2. a) GPE depends on mass, gravitational field strength, and height. [1 mark] The basketball has more mass, and height and gravitational field are the same, so its GPE is higher. [1 mark]

b) Kinetic store of the tennis ball [1 mark] and kinetic store of the basketball. [1 mark]

c) Elastic potential. [1 mark]

3. Bar chart with:
Before – energy in GPE store, [1 mark] labelled person pulled back. [1 mark]
After – energy in kinetic store, [1 mark] labelled person at bottom of swing. [1 mark]

4. Scientists did experiments [1 mark] and made measurements. [1 mark] They found that energy was conserved, [1 mark] but it is an abstract concept so took a long time to work out. [1 mark]

5. In both cases, it appears that energy has been created [1 mark] because it ends up higher than the distance it was dropped/deformed. [1 mark] In the tennis ball's case, the extra energy was from the basketball bouncing to a lower height. [1 mark] In this case, energy is transferred to an elastic store when you deform it [1 mark] which is returned to a gravity store when it bounces, [1 mark] so it ends up at a bigger height. [1 mark]

Physics Chapter 2.3: Apply: Energy transfer

Summary questions

1. [4 marks, 1 mark per correct answer]
radiation – transfer from the Sun
electric current – transfer from a power station
heating – transfer through clothing
forces – transfer when falling

2. a) Energy is transferred from the air through the block. [1 mark]

b) The material has a low thermal conductivity, [1 mark] so energy is transferred too slowly to melt the ice. [1 mark]

c) Energy transferred from the air and through the block will eventually melt the ice. [1 mark]

3. The handlebar has a high thermal conductivity, [1 mark] so energy is transferred quickly away from your hand and it feels cold. [1 mark] The handlebar grip has a low thermal conductivity, [1 mark] so energy is transferred slowly away from your hand and it feels warm. [1 mark]

4. a) Thickness of insulation, [1 mark] time in the hot/cold, [1 mark] surface area of block that is covered. [1 mark]

b) A sudden temperature change due to opening of door. [1 mark]

c) The thermometer does not read zero at the freezing point of water. [1 mark]

5. Make a model of a capsule with a thermometer [1 mark] and cover the capsule with insulation. [1 mark] Subject the insulation to extreme heat, for example, from a Bunsen burner, [1 mark] and measure the temperature inside. [1 mark] Repeat with other materials [1 mark] and choose the material for which the temperature inside does not increase/increases the least. [1 mark]

Physics Chapter 2.4: Apply: Energy resources

Summary questions

1. Coal, oil, gas, renewable, chemical, thermal [6 marks]

2. A fuel is burned to turn water to steam [1 mark] which drives a turbine. [1 mark] This drives a generator [1 mark] which produces electricity. [1 mark]

3. For example, advantages - does not take up extra space, [1 mark] produces electricity without burning fuels [1 mark] disadvantages - could get dirty and not work, [1 mark] could get damaged by cars/people. [1 mark]
4. a) Maintaining the p.d. in electrical appliances requires energy to be transferred from a chemical store in fuel to a chemical store in the appliance. [1 mark] To transfer this energy, the fuel is burned, which contributes to climate change. [1 mark]
   b) Electric cars do not burn petrol, so they do not produce CO_2 when they are running. [1 mark] This means that they do not contribute to climate change while running. [1 mark] They need to be charged, which transfers energy from a resource. [1 mark] If that involves burning a fossil fuel in a power station, then it still contributes to climate change. [1 mark]
5. How much land would be needed [1 mark] because it could be used for other things. [1 mark] How the power stations would work [1 mark] because they could be designed for other fuels. [1 mark] Whether the biofuels would be more expensive [1 mark] since people would not use them. [1 mark]

Physics Chapter 2.5: Apply: Conservation and dissipation

Summary questions

1. [4 marks, 1 mark per correct answer]
   dissipation – not useful transfer
   conservation – keeping the same
   lubrication – reduces friction
   insulation – reduces transfer by heating
2. [1 mark – more energy in chemical before than after, 1 mark – more energy in thermal store after than before, 1 mark – at least one bar of energy in all stores]

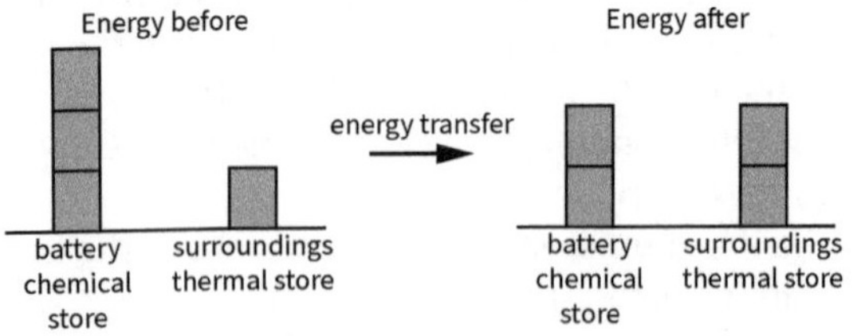

3. a) The same, [1 mark] a battery is still powering a lamp. [1 mark]
   b) A current flows in the wires, heating the torch. [1 mark]
   c) Use more conductive wires [1 mark] and a weaker power source. [1 mark]
4. Same energy in the thermal store of each house at the end, [1 mark] but more energy in the thermal store of the surroundings for the uninsulated house at the end [1 mark] and less energy in the chemical store of the coal for the insulated house. [1 mark]
5. Combined heat and power stations divert some dissipated energy to heat local houses. [1 mark] The houses will need to use less energy to be heated, [1 mark] so less fossil fuel would need to be burned. [1 mark] Burning fossil fuels produces carbon dioxide [1 mark] which fuels climate change, including sea level rise, [1 mark] so sea level rise would be less. [1 mark]

Physics Chapter 2.6: Apply: Work and power

Maths box

750 000 W = 750 kW = 0.75 MW

Summary questions

1. [4 marks, 1 mark per correct answer]
   work – force × distance
   power – energy/time
   unit of work – joule/newton metre
   unit of power – watt
2. a) Power = energy/time = 3000 J/4 s [1 mark]
   = 750 W [1 mark answer, 1 mark unit]
   b) 750 W = 0.75 kW [1 mark]
3. a) 12 kW = 12 kW × 1000 W/kW [1 mark]
   = 12 000 W [1 mark]
   1.5 hours = 1.5 × 3600 s/h [1 mark]
   = 5400 s [1 mark]
   b) Energy = power × time = 12 000 W × 5400 s [1 mark]
   = 64 800 000 J [1 mark answer, 1 mark unit]
4. If 20% of power is transferred as heat and sound, then only 80% of power is available for lifting.
   Available power to lift the car = 0.8 × 20 kW = 16 kW [1 mark]
   = 16 000 W [1 mark]
   Energy = force × distance = 8000 N × 32 m = 256 000 J [1 mark]
   Energy = power × time, so time = energy/power [1 mark]
   = 256 000 J/16 000 W [1 mark]
   = 16 s [1 mark]

Physics Chapter 2.7: Reach: Efficiency

Summary questions

1. Output, input, more, less [4 marks]
2. a) A high proportion of the total energy supplied is transferred into useful energy. [1 mark]
   b) $\text{Efficiency} = \frac{\text{useful output energy transfer (J)}}{\text{total input energy transfer (J)}} \times 100\%$
   $\text{Efficiency} = \frac{2000\text{ J}}{4000\text{ J}} \times 100\%$ [1 mark]
   = 50% [1 mark answer, 1 mark unit (%)]
3. a) Output energy = efficiency × input energy [1 mark]
   = 0.4 × 1500 J [1 mark]
   = 600 J [1 mark answer, 1 mark unit]
   b) Efficiency = 0.7 [1 mark]
   Useful power output = efficiency × total power output
   = 0.7 × 100 W [1 mark]
   = 70 W [1 mark answer, 1 mark unit]
4. Less efficient light bulbs would heat the house as well as light it, [1 mark] so they would burn out quickly in cold temperatures. [1 mark] Energy saving light bulbs transfer less energy by heating, so last longer in cold temperatures. [1 mark]

Physics Chapter 2.8: Reach: Elastic energy and Hooke's law

Summary questions

1. Work, elastic, extension, spring constant [4 marks]
2. a) A larger extension means the distance is larger. [1 mark] The work done depends on force and distance, so more work is done. [1 mark]
   b) $E = 0.5\,kx^2 = 0.5 \times 50$ N/m x $(0.02\text{ m})^2$ [1 mark change cm to m, 1 mark numbers]
   = 0.01 J (or 0.01 Nm) [1 mark answer, 1 mark unit]

a) $k = F/x = 8$ N/0.002 m [1 mark numbers]
   = 4000 N/m [1 mark numbers, 1 mark unit]
   $E = 0.5\,kx^2 = 0.5 \times 4000\text{N/m} \times (0.002\text{ m})^2$ [1 mark numbers]
   = 0.008 J (or 0.008 Nm) [1 mark answer, 1 mark unit]
   b) Area = 0.5 × 8 N × 0.002 m [1 mark numbers]
   = 0.008 J [1 mark answer, 1 mark unit]

Physics Chapter 2 Summary Answers

1. Elastic, thermal, nuclear [1 mark for all three]
2. a) Work = force × distance = 500 N × 100 m [1 mark]
   = 50 000 Nm [1 mark value, 1 mark unit]
   b) C – the energy transferred to a store we don't need or want [1 mark]
   c) Thermal. [1 mark]
3. a) It produces carbon dioxide. [1 mark]
   b) Use more efficient devices. [1 mark] and use devices for less time. [1 mark]
4. a) 'How does the distance you compress the spring affect how high it jumps?' [1 mark]
   Independent – distance compressed [1 mark]
   Dependent – height it jumps [1 mark]
   Control – type of spring, type of rod [2 marks]
   b) Appropriate start and end points, e.g., compressed and at the top of the jump [1 mark]
   Correct stores, e.g., elastic, gravity, thermal [1 mark]
   Correct energy proportions [1 mark – all elastic & no gravity before, 1 mark – some thermal before, 1 mark – no elastic & some gravity after, 1 mark – more in thermal store after]

Energy before-spring compressed			Energy after-spring at top of jump		
Elastic store	Gravity store	Thermal store	Elastic store	Gravity store	Thermal store
Compressed spring	Spring at the bottom	Surroundings	Uncompressed spring	Spring at the top	Surroundings

c) Use a lubricated spring. [1 mark]

Energy before – rollercoaster at top			Energy after – rollercoaster at bottom		
Gravity store	Kinetic store	Thermal store	Gravity store	Kinetic store	Thermal store

5. a) Kinetic store, [1 mark] thermal store [1 mark]
   b) One box coloured on the thermal store, [1 mark] 3 boxes coloured on the kinetic store. [1 mark]
   c) Work, [1 mark] heating by particles. [1 mark]
   d) Energy is not something that you experience/is only something mathematical. [1 mark]
   e) Power = energy/time [1 mark]
   = 600 000 J/20 s [1 mark]
   = 30 000 W [1 mark number, 1 mark unit]
6. a) $\text{Efficiency} = \frac{\text{useful output energy transfer}}{\text{total input energy transfer}} \times 100\%$ [1 mark]

   $\text{Efficiency} = \frac{35\text{ J}}{50\text{ J}} \times 100\%$ [1 mark]
   = 70% [1 mark]
   b) $\text{Efficiency} = \frac{\text{useful output energy transfer}}{\text{total input energy transfer}}$ [1 mark]
   Useful output power transfer = efficiency × total input power transfer [1 mark]
   = 0.45 × 10 000 W [1 mark]
   = 4500 W [1 mark]
7. From the graph: F = 20 N, x = 200 x 10^{-2} m (or 0.2 m) [1 mark for both]
   E = 0.5 F x [1 mark]
   = 0.5 × 20 × 0.2 m [1 mark]
   = 2 J (or 2 Nm) [1 mark]

Physics Chapter 3: Waves, sound, and light

Physics Chapter 3 Opener

Reactivate your knowledge questions

1. Only green light is allowed through and all other colours are absorbed
2.

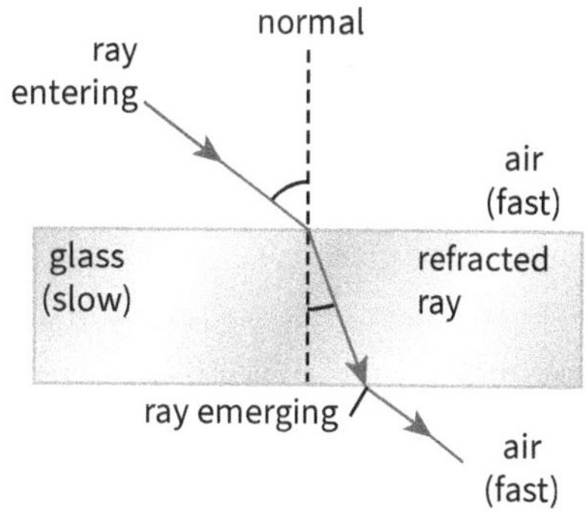

3. When you mix the primary colours of light, you get white; when you mix the primary colours of paint, you get black.

Physics Chapter 3.2: Apply: Wave properties

Summary questions

1. Transverse, S-wave, frequency, Hertz [4 marks]
2. a) Wavelength is the distance between two peaks on a snapshot. [1 mark] Time period is the time for one wave on a time trace. [1 mark]

   b) P-waves are longitudinal, S-waves are transverse. [1 mark] P-waves are primary waves, S-waves are secondary. [1 mark]
3. Sound waves are longitudinal and consist of areas of high and low pressure. [1 mark] When the sound wave hits the microphone, it produces an electrical signal. [1 mark] The amplitude of the signal depends on the pressure in the sound waves. [1 mark] Areas of high pressure, compressions, produce a peak [1 mark] and areas of low pressure, rarefactions, produce a trough. (1 mark] This produces a transverse wave. [1 mark]
4. a) 4 m [1 mark]

   b) Frequency doubles and time period halves, [1 mark] so 2 s. [1 mark]

   c) 0.01 s [1 mark]

Physics Chapter 3.3: Apply: Sound and its applications

Summary questions

1. Frequency, ultrasound, speed, seismic [4 marks]
2. a) Time = 0.02/2 [1 mark] = 0.01 s [1 mark]

   b) $d = s \times t$ = 3500 m/s × 0.01 [1 mark]
   = 35 m [1 mark answer, 1 mark unit]
3. Earthquakes produces longitudinal P-waves [1 mark] and transverse S-waves. [1 mark] They are mechanical because they need a medium to travel through. [1 mark] Seismometers on the other side of the Earth to the earthquake detect P-waves, [1 mark] but do not detect S-waves. [1 mark] S-waves don't travel through liquids so part of the core must be liquid. [1 mark]
4. a) 4 squares in time = 0.1 ms × 4 = 4×10^{-4} s [1 mark]
   Time for pulse to reach the back of the eye = 4×10^{-4} s/2 [1 mark]
   = 2×10^{-4} s [1 mark]
   $d = s \times t$ = 1500 m/s × 2×10^{-4} s [1 mark]
   = 0.3 m [1 mark answer, 1 mark unit]

   b) The time is ¼ the time for the reflection to the back of the eyeball, [1 mark] so the distance is ¼ or 0.3 m/4 (= 0.075 m). [1 mark]

Physics Chapter 3.4: Apply: Reflection and refraction

Summary questions

1. Reflects, angles, refracts, focus, ray, images, virtual, real [8 marks]
2. a) You have to trace the rays back to the image. [1 mark]

   b) The rays entering the eye would become more parallel [1 mark] and the image would become smaller. [1 mark]
3. a) The rays meet at the image. [1 mark]

   b) The image would be drawn [1 mark] bigger. [1 mark] The ray diagram would have a ray through the centre of the lens [1 mark] and a ray through the focus [1 mark] meeting at the top of the image. [1 mark]
4. In Figure 6, the object is between the focus and the lens. [1 mark] In Figure 5, the object is beyond the focus. [1 mark] When you put the glass close to the arrow in Figure 3, the arrow is between the focus and the glass, so it is like Figure 6. [1 mark] As you move it away there is a point where the arrow is beyond the focus, so it is like Figure 5. [1 mark]

Physics Chapter 3.5: Apply: Applications of reflection and refraction

Summary questions

1. [All correct – 4 marks, 3 correct – 3 marks, 2 correct – 2 marks, 1 correct – 1 mark]
   eyepiece lens – lens that magnifies the image
   reflecting telescope – telescope with mirrors
   concave mirror – mirror in a telescope
   objective lens – lens near the object
   refraction telescope – telescope with lenses
2. a) Eyepiece. [1 mark]

   b) Real. [1 mark]
3. a) The lenses disperse light (refract different colours of light by different amounts), but mirrors do not. [1 mark]

   b) [1 mark – second ray through focus, 1 mark – rays traced down, 1 mark – image drawn]

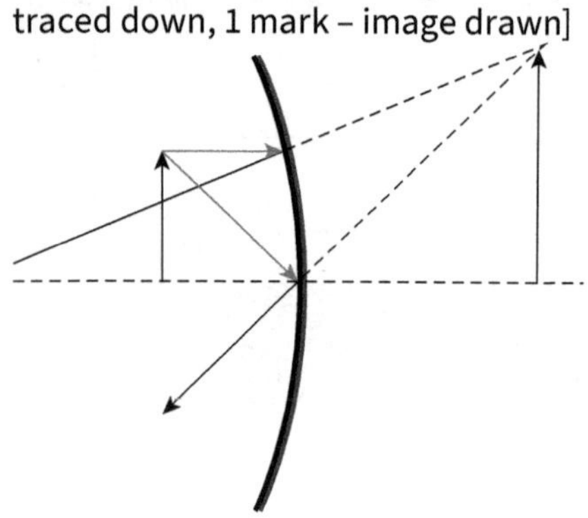

Physics Chapter 3.6: Apply: Light and colour

Summary questions

1. [All correct – 4 marks, 3 correct – 2 marks, 2 correct – 2 marks, 1 correct – 1 mark]
   primary colours of light – are red, green, blue
   white light – is made of 7 colours
   secondary colours – are yellow, cyan, magenta
   cone cells – detect primary colours
   rod cells – do not detect colour
2. **a)** Green and blue. [1 mark]
   b) Red and green. [1 mark]
   c) No cells are activated. [1 mark]
3. When you mix the primary colours of light, you get the white light, [1 mark] so the colours add. [1 mark] When you add paint colours together you get black, [1 mark] so the colours subtract. [1 mark]

Physics Chapter 3.7: Reach: Electromagnetic spectrum

Summary questions

1. Rays, smallest, ultraviolet, X-rays, gamma, cancer, heating [7 marks]
2. Gamma rays, X-rays, ultraviolet, visible, infrared, microwaves, radio waves [all correct – 3 marks, 5 – correct 2 marks, 3 or 4 correct – 1 mark]
3. **a)** Microwaves, [1 mark] X-rays, [1 mark] gamma rays. [1 mark]
   b) Lots of humans would get cancer [1 mark] because X-rays and gamma rays are ionising. [1 mark]

Physics Chapter 3.8: Reach: Applications of electromagnetic waves

Summary questions

1. Infrared, microwaves, matter [3 marks]
2. Radio – communication [1 mark] Microwaves – mobile phones [1 mark] Infrared – remote controls [1 mark] Visible – seeing [1 mark] Ultraviolet – detecting forgeries [1 mark] X-rays – security at airports [1 mark] Gamma – radiotherapy [1 mark]
3. **a)** A pulse of radio waves sent out and reflected back. [1 mark] The distance is calculated using speed × half the time of the return of the pulse. [1 mark]
   b) It is not ionising, [1 mark] so would not damage/kill cancer cells. [1 mark]
   Accept other suitable examples.
4. For example, the host speaks into a microphone [1 mark] which produces an electrical signal. [1 mark] An oscillator produces a microwave carrier wave [1 mark] and the electrical signal is mixed with the carrier wave and sent from an aerial. [1 mark] The phone picks up the carrier wave and extracts the sound signal, [1 mark] and the signal operates the loudspeaker in your phone. [1 mark]

Physics Chapter 3 Summary Answers

1. **a)** Transverse [1 mark]
   b) S-waves [1 mark]
   c) Wavelength [1 mark]
   d) Frequency, [1 mark] amplitude. [1 mark]
2. **a)** Reflection of sound [1 mark]
   b) B – Divide the time for the echo by two and multiply by the speed of sound in the material. [1 mark]
3. **a)** There was no communication at that time between distant countries/Al Burini spoke a different language. [1 mark]
   b) Magenta is made by mixing red and blue light. [1 mark] This is perceived when red and blue cones are activated in the retina, [1 mark] but it does not have its own spectral frequency. [1 mark]
   c) A colour of paint reflects the colour it is made from [1 mark] and absorbs the rest. [1 mark] If you mix all the primary colours of paint it appears black [1 mark] because all colours of light will be absorbed. (1 mark) If you mix all the primary colours of light you get white light [1 mark].
4. **a)** Line extrapolated back [1 mark] and image of fish drawn and labelled. [1 mark]

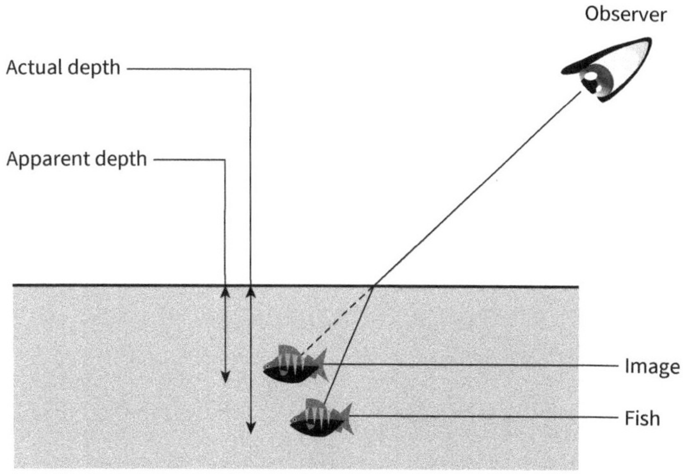

   b) If birds dived at an angle, they would miss the fish [1 mark] since they see the fish as above where it actually is. [1 mark]
5. **a)** The chlorophyll absorbs most of the violet/blue and red light. [1 mark] The green light activates the green cone cells in the retina [1 mark] and we perceive the leaf as being 'green'. [1 mark]
   b) The carotenoids absorb violet and most of blue. [1 mark] The red and green cones are activated, and blue but to a smaller extent, [1 mark] so the mixture is perceived as orange. [1 mark]
6. **a)** Radio waves, [1 mark] microwaves, [1 mark] infrared, [1 mark] visible, [1 mark] ultraviolet, [1 mark] X-rays, [1 mark] and gamma. [1 mark] Only ultraviolet, X-rays, and gamma are ionising.
   b) Ionising radiation can cause mutations in the DNA of cells [1 mark], which can cause mutations/cancer. [1 mark]
   c) The waves are in order of increasing frequency/decreasing wavelength. [1 mark]

7. a) A patient is injected with material that emits gamma rays. [1 mark] The material is taken up by particular organs [1 mark] and a gamma camera detects the emitted gamma rays. [1 mark]
   b) The camera can detect whether the organs are functioning normally [1 mark] and the risk of cancer is worth the diagnosis. [1 mark]

Physics Chapter 4: Electricity and magnetism

Physics Chapter 4 Opener

Reactivate your knowledge questions

1. 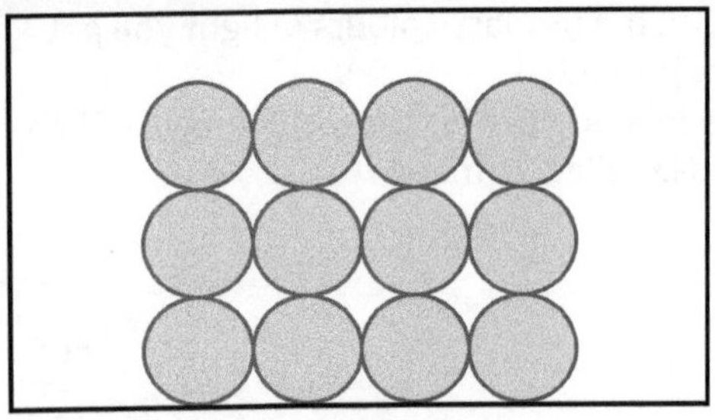

   The circles represent particles
2. The rope would be used up
3. A positive charge repels another positive charge, not the north pole of a magnet.

Physics Chapter 4.2: Apply: Static electricity and charge

Summary questions

1. [All correct – 3 marks, 2 correct – 2 marks, 1 correct – 1 mark]
   earthing – removing charge from an object
   friction – how some objects become charged
   negative – charge on an electron
   spark – what you get when the air conducts
2. a) There is not a big enough p.d. [1 mark]
   b) If there is a large p.d., the air conducts. [1 mark] The current produces light and heats the air to produce sound. [1 mark]
3. For example, petrol or aircraft fuel flowing in a pipe. [1 mark]
4. a) Rubbing it [1 mark] transfers electrons from the can to the cloth. [1 mark]
   b) The ground. [1 mark]
5. When lightning from a cloud strikes the tower, electrons transfer from the cloud to the tower [1 mark] through the air with a large p.d. [1 mark] The copper wire has low resistance [1 mark], so the electrons can easily flow through it to the ground. [1 mark] This removes the excess electrons from the tower, [1 mark] making the tower neutral. [1 mark]

Physics Chapter 4.3: Apply: Current, p.d., and resistance

Summary questions

1. Electric, charged, opposite, gripping [4 marks]
2. a) The pipe would be wider, [1 mark] so the water can flow more easily. [1 mark]
   b) The pump goes faster, [1 mark] so there is a bigger push on the water. [1 mark]
3. a) Conventional current goes from positive to negative. [1 mark] Connecting the negative connection on the ammeter to the positive terminal on the battery produces a negative reading. [1 mark]
   b) Connecting the battery produces an electric field. [1 mark] Electrons are charged particles, [1 mark] so there is a force from the electric field on the electrons. [1 mark] This unbalanced forces causes them to accelerate and move around the circuit. [1 mark]

Physics Chapter 4.4: Apply: Series and parallel

Summary questions

1. Variable, dimmer, stay on, parallel [4 marks]
2. a) Circuit as in Figure 4. [2 marks, with 1 mark deducted for each wrong feature]
   b) The current in the circuit. [1 mark]
   c) In the winter, there would be less light, so the resistance of the LDR would be higher, [1 mark] so the current in the circuit would be smaller. [1 mark]
3. a) R = V/I = 240 V/1 A [1 mark]
   = 240 Ω [1 mark number, 1 mark unit]
   b) R = V/I = 240 V/4 A [1 mark]
   = 60 Ω [1 mark number, 1 mark unit]
   c) The appliance might overheat/melt the casing. [1 mark]

Physics Chapter 4.5: Apply: Magnetism

Summary questions

1. Domains, same, random, stroking, hard [5 marks]
2. a) An induced magnet is a magnetic material that becomes a magnet when it placed in a magnetic field and will be weakly magnetic or non-magnetic when the field is removed. [1 mark] A permanent magnet has been places in a field long enough to remain magnetic when the field is removed. [1 mark]
   b) In steel, the domains line up when it is in the field and remain lined up with the field is removed. [1 mark] In iron, they do not stay lined up with the field is removed. [1 mark]

3. a) [1 mark]

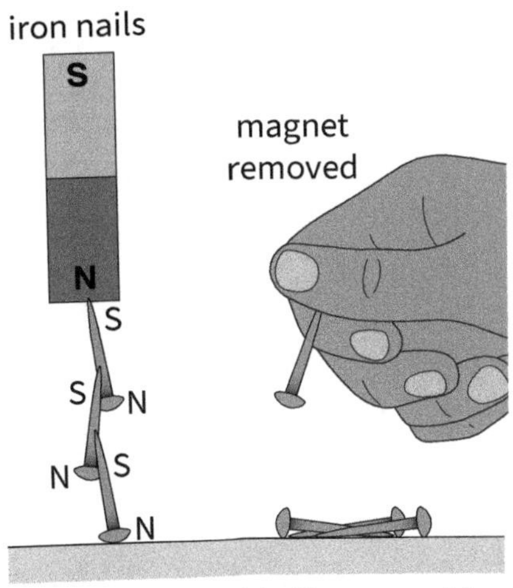

b) If the nail closest to the magnet is attracted, it must have the south pole in contact with the north pole of the magnet, [1 mark] which makes the other end of that nail a north pole. [1 mark] All the other nails will be similarly polarised. [1 mark]

4. a) If the domains are lined up inside a piece of material, then when it is cut in half, the domains will still line up. [1 mark] This makes any smaller piece into a magnet with the same orientation of poles. [1 mark]

b) The domains are moved one way and then another so will not stay completely aligned. [1 mark]

c) The Earth's magnetic field is not strong enough [1 mark] to align enough domains to magnetise it. [1 mark]

Physics Chapter 4.6: Reach: Electromagnetism and induction

Summary questions

1. P.d., magnet, pole [3 marks]
2. a) Move the north pole of the magnet in first (to the right), [1 mark] move the coil towards the magnet (to the left). [1 mark]

   b) Move the magnet/coil faster [1 mark] or use stronger magnets. [1 mark]
3. a) A coil of wire spins [1 mark] between the poles of two magnets (that are attracting). [1 mark]

   b) For example, you can charge your phone anywhere. [1 mark]

Physics Chapter 4.7: Reach: Alternating current and the National Grid

Summary questions

1. Alternating, direct, magnetic field, potential difference. National Grid [5 marks]
2. a) Any three from: power station, step-up transformer, pylons, step-down transformer, underground transmission lines [3 marks]

   b) Step-up transformers increase the p.d. [1 mark] and step-down transformers decrease the p.d. [1 mark]
3. The changing current in the coil of the induction coil produces a changing magnetic field. [1 mark] The metal pan is a conductor, [1 mark] so a p.d. is induced in the metal pan. [1 mark] This produces currents in the metal pan, because there is a complete circuit, (1 mark] which heat the pan and the food. [1 mark] The ceramic top is an insulator, so no currents flow. [1 mark]
4. Steel is magnetic, but aluminium is not. [1 mark] The magnetic field will be stronger in the steel than the aluminium, [1 mark] so the p.d. will be bigger/you need a smaller current in the induction coil to heat the pan. [1 mark]

Physics Chapter 4 Summary Answers

1. Battery, more, potential, bigger, field [5 marks]
2. a) Diagram with correct shape of field lines [1 mark] and arrows from north to south. [1 mark]

   b) The domains in the iron line up in the magnetic field of the magnet. [1 mark] There is no field around a piece of iron, so the domains do not line up. [1 mark]
3. a) B [1 mark] since there is more than one loop. [1 mark]

   b) In a series circuit, the resistances add up (so 4 Ω + 4 Ω = 8 Ω). [1 mark]

   c) Current = p.d./resistance [1 mark]
   = 12 V/8 Ω [1 mark]
   = 1.5 A [1 mark]

   d) It would increase [1 mark] since the total resistance of the circuit would decrease (now there is more than one loop). [1 mark]

   e) Add a light dependent resistor instead of one of the resistors. [1 mark] Put a voltmeter across either component and it will change with the light level. [1 mark]
4. a) They both have the same resistance [1 mark] and the same p.d./p.d. of the battery. [1 mark]

   b) Current in one resistor = p.d./resistance [1 mark]
   = 12 V/4 Ω [1 mark]
   = 3 A [1 mark]
   Total current = 6 A [1 mark]

   c) Resistance is less when the switch is closed. [1 mark] The p.d. is the same, but the current is bigger. [1 mark]

   d) Toaster/oven with different hot plates. [1 mark]
5. a) For example, the rope model does not show a field while the gravitational field model does. [1 mark]

   b) For example, the rope model clearly shows resistance [1 mark] while the gravitational field model does not show it so well. [1 mark]
6. a) When a current flows, the coil becomes an electromagnet/magnetic [1 mark] and there is a force because the electromagnetic is repelled/attracted by the permanent magnet. [1 mark]

   b) Spin the coil [1 mark], disconnect the battery, and connect a voltmeter instead. [1 mark]

   c) Use stronger permanent magnets, [1 mark] spin it faster, [1 mark] and use more turns of wire on the coil. [1 mark]
7. a) The toaster works on **alternating** current. [1 mark]

   b) The **potential difference** is generated in a power station and transferred by the National Grid. [1 mark]

   c) The p.d. is increased by a **transformer** in the National Grid. [1 mark]

OXFORD
UNIVERSITY PRESS

Great Clarendon Street, Oxford, OX2 6DP, United Kingdom

Oxford University Press is a department of the University of Oxford. It furthers the University's objective of excellence in research, scholarship, and education by publishing worldwide. Oxford is a registered trade mark of Oxford University Press in the UK and in certain other countries.

First published in 2022

British Library Cataloguing in Publication Data
Data available

978-1-38-202111-1

978-1-38-202099-2 (ebook)

10 9 8 7 6 5 4 3 2 1

Paper used in the production of this book is a natural, recyclable product made from wood grown in sustainable forests.

The manufacturing process conforms to the environmental regulations of the country of origin.

Printed and bound by CPI group (UK) Ltd, Croydon, CRO 4YY

Acknowledgements
The authors would like to thank the following:

Jo Locke:
Many thanks to my girls Emily and Hermione for all their support, encouragement, and delicious cakes, and to Dave for providing endless cups of tea and helpful ideas. It has been an honour to write again with the original Activate team of Philippa Gardom Hulme and Helen Reynolds; I couldn't have done it without you.

Alyssa Fox-Charles:
Thanks to Billy and my ever-supporting family Judy, Nic, Gemma, and Josie for their encouragement, help, and never-ending belief in me, and OUP for allowing me this amazing opportunity.

The publisher and authors would like to thank the following for permission to use photographs and other copyright material:

Cover by Michal Bednarski.

Artwork by Q2A Media, Michal Bednarski, Phoenix Photosetting, and Oxford University Press.